Parvoviruses and Human Disease

Editor

John R. Pattison, FRC Path.
Professor
Department of Medical Microbiology
University College and Middlesex
School of Medicine
London, England

CRC Press
Taylor & Francis Group
Boca Raton London New York

CRC Press is an imprint of the
Taylor & Francis Group, an informa business

First published 1988 by CRC Press
Taylor & Francis Group
6000 Broken Sound Parkway NW, Suite 300
Boca Raton, FL 33487-2742

Reissued 2018 by CRC Press

Library of Congress Cataloging-in-Publication Data

Parvoviruses and human disease.

Includes bibliographies and index.
1. Parvovirus infections. 2. Parvoviruses.
I. Pattison, J. R. (John Ridley) [DNLM: 1. Parvoviridae.
2. Parvovirus Infections. QW 165.5.P3 P2755]
QR201.P33P37 1988 576'.6484 87-21840
ISBN 0-8493-5956-2

A Library of Congress record exists under LC control number: 87021840

Publisher's Note
The publisher has gone to great lengths to ensure the quality of this reprint but points out that some imperfections in the original copies may be apparent.

Disclaimer
The publisher has made every effort to trace copyright holders and welcomes correspondence from those they have been unable to contact.

ISBN 13: 978-1-315-89630-4 (hbk)
ISBN 13: 978-1-351-07540-4 (ebk)

Visit the Taylor & Francis Web site at http://www.taylorandfrancis.com and the
CRC Press Web site at http://www.crcpress.com

PREFACE

Human parvovirus B19 was first discovered in the 1970s as an incidental finding during screening for hepatitis B virus. At first its pathogenic significance was not appreciated, but during the 1980s a series of diseases have been shown to be due to B19 virus. This book gives details of the discovery and research work on B19 virus to date. The virus is an autonomous parvovirus and many of its properties and much of its behavior can be predicted from this. Accordingly, the detailed account of B19 virus is set in the context of two general chapters on the nature of parvoviruses and the patterns of disease in animals produced by parvoviruses. In human medicine the study of virus infections depends greatly on accurate diagnostic tests. This aspect is covered in a chapter prior to a detailed discussion of the two principal clinical manifestations of B19 virus infection, namely, aplastic crisis and a rash illness often associated with arthralgia. Volunteer studies have subsequently provided a unifying hypothesis for B19 virus induced disease and laboratory studies have provided a basis for pathogenicity at the cellular level. Finally, with B19 virus the unanswered question of what effect the infection has when it occurs in pregnancy is addressed. B19 virus has been intensively researched during the last 8 years, but it is not the only human parvovirus. The two final chapters review the possible role of parvoviruses in gastroenteritis in humans and whether adeno-associated viruses have any causal role in disease.

Parvoviruses are versatile and wide-ranging pathogens in animals. It may well be that parvoviruses hold further surprises in terms of human disease. However so much progress has been made recently in our understanding of this subject that a volume detailing current knowledge seems very timely.

John R. Pattison
London, England

THE EDITOR

John R. Pattison, M.A., D.M., FRC Path., is professor of medical microbiology in the University College and Middlesex School of Medicine of the University of London.

Dr. Pattison qualified in medicine in 1968, having received training at the University of Oxford and the Middlesex Hospital Medical School. During that time he had already spent a year doing research in microbiology and after qualifying, he immediately pursued a career in medical microbiology. His first senior staff appointment was in virology jointly at the London Hospital Medical College and St. Bartholomew's Hospital Medical School. He was appointed professor of medical microbiology at Kings College Hospital Medical School London in 1977 and took up his present appointment in 1984.

Dr. Pattison's special interest is clinical virology. He has done research work into aspects of rubella virus infection. This brought congenital abnormalities into focus for him and it was for this reason that he first started work on human parvovirus in 1979.

CONTRIBUTORS

Mary J. Anderson, Ph.D.
Senior Lecturer
Department of Medical Microbiology
University College and Middlesex School of Medicine
London, England

Neil R. Blacklow, M.D.
Professor
Division of Infectious Diseases
University of Massachusetts Medical School
Worcester, Massachusetts

E. Owen Caul, Ph.D.,
Top Grade Microbiologist
Department of Virology
Public Health Laboratory
Bristol, England

Bernard J. Cohen, Ph.D.
Microbiologist
Virus Reference Laboratory
Central Public Health Laboratory
London, England

Susan F. Cotmore, Ph.D.
Associate Research Scientist
Department of Laboratory Medicine
Yale University Medical School
New Haven, Connecticut

Amanda Goldstein, MRCP
MRC Laboratories
University of West Indies
Kingston, Jamaica

Keiya Ozawa, Ph.D.
Visiting Fellow
Clinical Hematology Branch
National Heart, Lung, and Blood Institute
National Institutes of Health
Bethesda, Maryland

John R. Pattison, FRC Path.
Professor
Department of Medical Microbiology
University College and Middlesex School of Medicine
London, England

Graham R. Serjeant, FRCP
Director
MRC Laboratories
University of West Indies
Kingston, Jamaica

Günter O. Siegl, Dr. rer. nat.
Professor
Institute for Hygiene and Medical Microbiology
University of Bern
Bern, Switzerland

Peter Tattersall, Ph.D.
Associate Professor
Departments of Laboratory Medicine and Human Genetics
Yale University School of Medicine
New Haven, Connecticut

D. A. J. Tyrrell, FRC Path.
Director
MRC Common Cold Unit
Harvard Hospital, Salisbury
Wilts, England

Neal S. Young, M.D.
Chief
Clinical Hematology Branch
National Heart, Lung, and Blood Institute
National Institutes of Health
Bethesda, Maryland

TABLE OF CONTENTS

Chapter 1

THE DISCOVERY OF HUMAN PARVOVIRUSES

J. R. Pattison

TABLE OF CONTENTS

I. INTRODUCTION

Any current outline of the classification of viruses contains a well-defined family of small, single stranded, DNA viruses, the Parvoviridae. This is further subdivided into three distinct genera, the parvoviruses (viruses of vertebrates capable of autonomous replication), the dependoviruses (again viruses of vertebrates but requiring a helper virus for replication), and the densoviruses of insects. This well-ordered state of affairs tends to obscure the fact that the discovery of many of these viruses was more accidental or incidental than the result of deliberate searches. Certainly the discovery of human adeno-associated viruses was a chance finding arising from the examination of adenovirus stocks by electron microscopy. Furthermore, the discovery of an autonomous human parvovirus which causes systemic disease with blood-borne spread was pure serendipity (see below). Historically, however, it seems likely that the first autonomous human parvovirus to be described will prove to be the small round viruses in feces first described by Paver and colleagues in 1973.[1] This discovery was the result of a search for the then elusive agent of nonbacterial gastroenteritis. Even now, however, there is no formal proof that these are autonomous parvoviruses and the question of whether they are pathogenic is addressed in Chapter 10. Although they are morphologically similar to the parvoviruses found in human serum, they are antigenically distinct.[2] It seems likely then that there are two quite different human parvoviruses and it is the explosion of interest in the serum virus during the 1980s that is the reason for this book.

Another consistent feature of the early history of parvoviruses is the tendency for the viruses to be discovered before their associated diseases. The only significant exception to this is the feline parvovirus in the sense that the fatal disease, feline panleukopenia, had been known and investigated since the end of the 19th century, but the causative virus was not defined until the 1960s.[3] Even with the host range variants, mink enteritis virus, and canine parvovirus the respective diseases in mink and dogs were the stimulus to the search for and characterization of the viruses. With the human serum parvovirus, the more usual pattern of the incidental finding of a virus apparently not associated with specific disease was preserved. However, as with other parvoviruses (e.g., the porcine and bovine parvoviruses) found in this way, further work has shown that they are significant pathogens. Indeed, the pathogenic potential of the parvoviruses, the smallest and simplest of the known viruses, is remarkable and the subject of many of the following chapters.

II. THE DISCOVERY OF B19 VIRUS

In the mid-1970s, the second generation tests (reverse passive hemagglutination, RPHA, and radioimmunoassay, RIA) for hepatitis B surface antigen (HBsAg) were in the process of replacing the first generation tests (gel diffusion, electrophoresis, and complement fixation). Clearly, the sensitivity and specificity of the old and the new had to be compared, and in one study[4] they were compared on the 3219 sera received in a routine clinical laboratory during a period of 12 weeks. The primary comparison was between electrophoresis and one of the commercially available RPHA tests. By titration of two known positive sera, the latter was 12 and 400 times more sensitive than electrophoresis. In the group as a whole, RPHA increased the yield of positive results by one third (159 as opposed to 123 by electrophoresis). In spite of the dramatically greater sensitivity of RPHA, there were three sera which were positive by electrophoresis but negative by two RPHA tests and RIA. Two of these were noted to be positive by electron microscopy. It was felt at the time that these anomalous results were due to variations in reactivity of the various HBsAg antibodies used in the tests. Electrophoresis tests used blood donor sera selected because of high titer anti-HBs (but also possessing, incidentally, all the reactivities expected of adult human sera).

The RPHA and RIA used hyperimmune animal sera from sheep, horse, and guinea pig. Thus the concept arose that no test was capable of detecting all HBsAg positive specimens. One of the sera which reacted only in electrophoresis was included in a quality control panel for HBsAg tests; it was number 19 in panel B.

It was then realized that these anomalous results were due to an antigen that was distinct from HBsAg and the details of a series of 11 patients with this antigenemia were published.[5] The reactivity of the sera in electrophoresis tests for HBsAg was due simply to the fact that these tests used human sera, some of which contained antibody to the newly discovered virus. By contrast, the RPHA and RIA tests used hyperimmune animal sera which were monovalent with respect to antibodies to human viruses. Electron microscopy of the antigen positive sera showed viral particles with an average diameter of 23 nm characteristically displaying disrupted fragments and empty shells. These appearances were considered characteristic of parvoviruses and the buoyant density in $CsCl_2$ was shown to be in the parvovirus range. The new antigen showed no antigenic relationship with the four adeno-associated virus types or Kilham's rat virus, and no hemagglutination could be demonstrated with cells from a variety of species.

The 11 blood samples were from 9 healthy blood donors, a case of acute hepatitis, and a patient who had received a renal transplant 1 week earlier. This pattern to some extent reflects the nature of specimens tested for HBsAg, and testing of specimens from 34 other cases of hepatitis showed them all to be negative. Second samples from 4 of the original 11 positive individuals were antigen negative but antibody positive, with the interval between the two samples varying from 2 weeks to 6 months. Further serological tests showed that the virus infection was common (approximately 30% of adults had antibody to the virus). Thus it was that a new human parvovirus was discovered, an apparently common infection in the population not at first associated with any specific clinical illness.

III. THE NOMENCLATURE OF HUMAN PARVOVIRUS

The titles of the first publications[1,2,5] related to possible human parvoviruses — "A Small Virus in Human Faeces", "Comparison of Human Fecal and Serum Parvo-Like Viruses", and "Parvovirus-Like Particles in Human Sera" — reveal that the authors were suitably cautious about classification and nomenclature. Even as papers began to emerge on the diseases associated with the serum virus, there was still no knowledge of the nature of the viral genome and authors used deliberately vague terms such as parvovirus-like agent (PVLA)[6] and serum parvovirus-like virus (SPLV).[7] Meanwhile, the designation B19 had been used in a number of other publications.[8,9] Similar anomalous HBsAg to those described above had been found by the French in 1972 and in one of the Japanese blood transfusions centers in 1979. For these the terms "Aurillac" and "Nakatani" antigens, respectively, were used and these were shown to have similar serological reactions to B19.[10,11]

The first data on the nature of the genome of the new virus were published by Summers and colleagues in 1983.[12] The virion deoxyribonucleic acid (DNA) proved to be single-stranded DNA 5.5 kb in length. Interestingly, the virus packages complementary strands into separate virions with equal efficiency. This was at one time thought to be an exclusive property of the dependo- and densoviruses, but under certain conditions the parvovirus LuIII separately encapsidates complementary strands. Further studies[13,14] confirmed the nature of the genome, and DNA:DNA hybridization revealed no relationship with the human adeno-associated viruses but a distant relationship with the helper-independent parvoviruses of rodents.[14]

There is no doubt that the known properties of the virus discovered by Cossart and colleagues in 1975 allow definitive classification within the family Parvoviridae. Assignment to a genus is slightly more difficult since no cell-culture system is available for this virus and formal proof that a helper virus is not required needs such a culture system. Nevertheless all the evidence points to the inclusion of this virus in the genus *Parvovirus* and this is the recommendation of the Study Group on Parvoviridae of the International Committee on Taxonomy of Viruses (ICTV). In its third report[15] the study group discuss the naming of the new human parvovirus. Observation of the tradition of naming new autonomous parvoviruses would lead to the designation "human parvovirus". Inevitably the acronym "HPV" would be used and this would lead to confusion with human papillomaviruses. The Study Group recommends referring to the new parvovirus exclusively as B19 virus. This is a somewhat trivial name although it fits in with the names of some of the early autonomous parvoviruses (H1, Lu III, TVX). Also it rather begs the question, whatever happened to viruses B1 to B18, and will make the logical naming of human fecal parvovirus (should there prove to be any) an interesting exercise. Nevertheless, the designation B19 virus certainly ensures that there will be no confusion with other human viruses. It is a name which will achieve widespread usage and therefore it will be used throughout the book.

REFERENCES

1. **Paver, W. K., Caul, E. O., Ashley, C. R., and Clarke, S. K. R.,** A small virus in human faeces, *Lancet,* 1, 237, 1973.
2. **Paver, W. K. and Clarke, S. K. R.,** Comparison of human fecal and serum parvolike viruses, *J. Clin. Micro.,* 4, 67, 1976.
3. **Johnson, R. H. and Cruikshank, J. G.,** Problems in classification of feline panleucopenia virus, *Nature (London),* 212, 622, 1966.
4. **Vandervelde, E. M., Goffin, C., Megson, B., Mahmood, N., Porter, A., and Cossart, Y. E.,** User's guide to some new tests for hepatitis-B antigen, *Lancet,* 2, 1066, 1974.
5. **Cossart, Y. E., Cant, B., Field, A. M., and Widdows, D.,** Parvovirus-like particles in human sera, *Lancet,* 1, 72, 1975.
6. **Anderson, M. J., Davis, L. R., Hodgson, J., Jones, S. E., Murtaza, L., Pattison, J. R., Stroud, C. E., and White, J. M.,** Occurrence of infection with a parvovirus-like agent in children with sickle cell anaemia during a two-year period, *J. Clin. Pathol.,* 35, 744, 1982.
7. **Kelleher, J. H., Luban, N. L. C., Mortimer, P. P., and Kamimura, T.,** The human serum "parvovirus". A specific cause of aplastic crisis in hereditary spherocytosis, *J. Pediatr.,* 102, 720, 1983.
8. **Edwards, J. M. B., Kessel, I., Gardner, S. D., Eaton, B. R., Pollock, T. M., Fleck, D. G., Gibson, P., Woodroof, M., and Porter, A. D.,** A search for a characteristic illness in children with serological evidence of viral or toxoplasma infection, *J. Infect. Dis.,* 3, 316, 1981.
9. **Shneerson, J. M., Mortimer, P. P., and Vandervelde, E. M.,** Febrile illness due to a parvovirus, *Br. Med. J.,* 2, 1590, 1980.
10. **Courouce, A. M., Ferchal, F., Muller, A., Soulier, J. P., Morinet, F., Drouet, J., and Perol, Y.,** Human parvovirus infections in France, *Lancet,* 1, 160, 1984.
11. **Okachi, K., Mori, R., Miyazaki, M., Cohen, B. J., and Mortimer, P. P.,** Nakatani antigen and human parvovirus (B19), *Lancet,* 1, 160, 1984.
12. **Summers, J., Jones, S. E., and Anderson, M. J.,** Characterisation of the genome of the agent of erythrocyte aplasia permits its classification as a human parvovirus, *J. Gen. Virol.,* 64, 2527, 1983.
13. **Clewley, J. P.,** Biochemical characterization of a human parvovirus, *J. Gen. Virol.,* 65, 241, 1984.
14. **Cotmore, S. F. and Tattersall, P.,** Characterization and molecular cloning of a human parvovirus genome, *Science,* 226, 1161, 1984.
15. **Siegl, G., Bates, R. C., Berns, K. I., Carter, B. J., Kelly, D. C., Kurstak, E., and Tattersall, P.,** Characteristics and taxonomy of Parvoviridae, *Intervirology,* 23, 61, 1985.

Chapter 2

THE NATURE OF PARVOVIRUSES

Peter Tattersall and Susan F. Cotmore

TABLE OF CONTENTS

I. INTRODUCTION

The Parvoviridae, or parvovirus family, consists of a large number of physically and chemically similar viruses which infect many animal species.[1] Table 1 lists the members of this family currently accepted by the International Committee on the Taxonomy of Viruses, and some possible additional members. These agents are small, nonenveloped, icosahedral viruses approximately 20 nm in diameter, with a 5 kbase single-stranded deoxyribonucleic acid (DNA) genome. The parvoviruses of vertebrates are divided into two genera whose properties are compared in Table 2. Members of the adeno-associated virus (AAV) genus, now renamed the *Dependovirus* genus, are defective and entirely depend upon adenovirus or herpes virus for their replication.[2] In contrast, members of the autonomous parvovirus genus, are capable of productive replication without the aid of a helper virus in the majority of host cells studied to date.[2] Replication of the autonomous viruses is, however, dependent upon cellular functions expressed transiently during the S phase of the cell cycle.[3-7] Since the viruses cannot induce resting cells to enter the S phase,[4] it is therefore not surprising that viral replication is restricted to proliferating cells both in vitro[4] and in vivo.[8-11] The B19 agent is the first parvovirus associated with human disease, and therefore is the major subject of this book. Since all the evidence so far points to its being independent of any requirement for a helper virus, and as the adeno-associated viruses are dealt with in a separate chapter, we will concentrate in the following pages on describing the characteristics of the autonomously replicating parvoviruses of vertebrates, their structure, genetic strategy, and virus-host cell interactions.

Numerous studies over the past 20 years on the pathogenicity of autonomous parvoviruses have shown that they are predominantly teratogenic agents. In general, they cause fetal and neonatal abnormalities by destroying specific cell populations which are rapidly proliferating during the normal course of development.[8-13] The details of parvoviral pathogenesis, as it has been observed in both animal and human systems, will be dealt with in other chapters. However, it is pertinent here to summarize a few aspects of their natural history for which there are in vitro analogues, and for which explanations, at the molecular level, are beginning to emerge.

Although proliferative activity appears to be a prerequisite for target organs, it is clear that not all tissues which turn over rapidly are necessarily subject to attack by autonomous parvoviruses.[11] While most adult tissues are mitotically quiescent compared to those of the fetus and neonate, many, such as gut epithelium and the lymphopoietic system contain large numbers of cycling cells. One might expect these cells, which are essential for the host organism's well-being and survival, to be targets for parvovirus attack in the adult. The sparing of these adult tissues by the majority of autonomous parvoviruses is underlined by the existence of a small subset of parvoviruses, namely the feline panleukopenia/mink enteritis/canine parvovirus group and the Aleutian disease virus of mink, which frequently cause fatal disease in adult animals involving extensive destruction of gut epithelium and reticuloendothelial cells.[12,13]

Studies on the replication of autonomous parvoviruses in vitro, particularly with the minute virus of mice (MVM), have provided significant support for the hypothesis that lytic virus growth is modulated by developmentally regulated components operating in the host at the cellular level. Mohanty and Bachmann[14] have reported that actively dividing cells of the early mouse embryo are resistant to killing by MVM. Murine embryonal carcinoma cells, the stem cells of teratocarcinoma, are resistant to the prototype strain MVM(p), as are many of their differentiated derivatives.[15,16] However, when these cells are induced to differentiate in vitro they give rise to at least one differentiated cell type, resembling a fibroblast, which supports lytic MVM(p) replication.[16] These studies suggest that cell cycling, though nec-

Table 1
PARVOVIRUSES OF VERTEBRATES

Autonomous Subgroup

Accepted members	Acronym
Rat virus	RV
H-1 virus (rat)	H-1
RT virus (rat)	RT
TVX (unknown)	TVX
Minute virus of mice	MVM
LuIII virus (unknown)	LuIII
B19 (human)	B19
Porcine parvovirus	PPV
Bovine parvovirus	BPV
Feline parvovirus	FPV
Species host range variants:	
Mink enteritis virus	MEV
Canine parvovirus	CPV
Racoon parvovirus	RPV
Lapine parvovirus	LPV
Aleutian disease virus (mink)	ADV
Goose parvovirus	GPV
Possible Members	
Minute virus of canines	MVC
HB virus (human?)	HB
RA-1 (human)	RA-1

Dependovirus Subgroup

Adeno-associated type 1 (monkey)	AAV-1
Adeno-associated type 2 (human)	AAV-2
Adeno-associated type 3 (human)	AAV-3
Adeno-associated type 4 (monkey)	AAV-4
Adeno-associated type 5 (human)	AAV-5
Bovine adeno-associated virus	BAAV
Canine adeno-associated virus	CAAV
Avian adeno-associated virus	AAAV
Possible members	
Equine adeno-associated virus	EAAV
Ovine adeno-associated virus	OAAV

Parvoviruses of Invertebrates

Galleria densovirus	Galleria DNV
Junonia densovirus	Junonia DNV
Agraulis densovirus	Agraulis DNV
Bombyx densovirus	Bombyx DNV

Note: Where the virus name does not include it, the primary host species of each virus is indicated in parentheses.

essary, is not sufficient for the lytic replication of parvoviruses, and that the differentiated state of the host cell is of paramount importance.

The circumstances of the first isolations of autonomous parvoviruses, mostly from tumors, originally suggested that the viruses might have some causal association with neoplastic disease in their hosts. Indeed, after the original isolations by Kilham of rat virus (RV) and by Toolan of H-1, attempts to isolate these viruses from nonmalignant tissues by the same

Table 2
PARVOVIRUSES OF VERTEBRATES

Autonomous parvoviruses	Dependoviruses
Helper independent in cells of normal host species	Completely dependent upon adeno- or herpes coinfection for productive infection
Pathogenic in host species	May influence course of helper-induced disease — not themselves pathogenic
Inhibit viral, chemical and spontaneous transformation in vivo and in vitro	Inhibit transformation by helper virus in vivo and in vitro
Require host cell factor(s) expressed transiently during S-phase	Inhibit helper virus replication
Unable to induce resting cells to enter mitotic cycle	Inhibited by high helper input
Require developmentally regulated host factor(s)	Host cell requirements predominantly dictated by those of helper virus
Genome usually has unique termini	Genome always terminally redundant
Package predominantly minus sense DNA strand — a few exceptions	Packages equal numbers of plus and minus sense DNA strands, in separate virions

procedures proved unsuccessful.[17,18] Since that time, however, many studies on the long-term influence of parvovirus infection on the host animal have shown the opposite conclusion to pertain, that these viruses interfere with, and in some cases appreciably suppress, the formation of spontaneous and virus- or carcinogen-induced tumors in their hosts.[19-22] The possibility that parvoviral suppression of oncogenic transformation is not merely another reflection of their requirement of dividing cells of a particular differentiated phenotype is afforded by a recent study of MVM interference with SV40 transformation reported by Mousset and Rommelaere.[23] They isolated a spontaneous mutant of Balb/c 3T3 mouse fibroblasts which was resistant to cell killing by MVM(p) by virtue of an intracellular block to virus replication. Although this cell line was identical to its MVM sensitive parent in its susceptibility to SV40 transformation, coinfection or superinfection of these mutant cells with MVM drastically reduced the ability of SV40 to transform them. Futhermore, the mutant cells regained the MVM-sensitive phenotype upon transformation with SV40.

Since differences in pathogenic potential exist not only between virus serotypes, but between virus strains of the same serotype,[24-28] this suggests that a particular tissue tropism might not be an invariant property of each virus. The isolation of an additional strain of minute virus of mice, MVM(i), as an immunosuppressive agent from a murine lymphoma indicates that a mutable genetic component in the virus may play a role in determining the type of differentiated cell the virus can lytically infect. MVM(i) suppresses a number of T-cell mediated functions as measured in vitro whereas MVM(p) does not,[28] despite their genomes being closely related in sequence. The molecular basis of parvovirus strain-dependent target cell specificity is now understood to some extent, and will be discussed later, in Section V.

II. STRUCTURE OF THE VIRION

A. Morphology

Parvoviruses are isometric, nonenveloped particles, 20 to 25 nm in diameter, which contain a single-stranded DNA genome of around 5000 nucleotides. Electron micrographs show icosahedral particles made up of multiple capsomers, which are easily distinguished from those of most other virus groups by their very small size (Figure 1A). Infectious particles resolve in isopycnic cesium chloride gradients into two forms: a major species with a buoyant density of 1.41 g/cm^3 (termed "light fulls") and a minor species with a density of 1.45 to 1.47 g/cm^3 (heavy fulls) which is probably an infectious precursor form of the mature

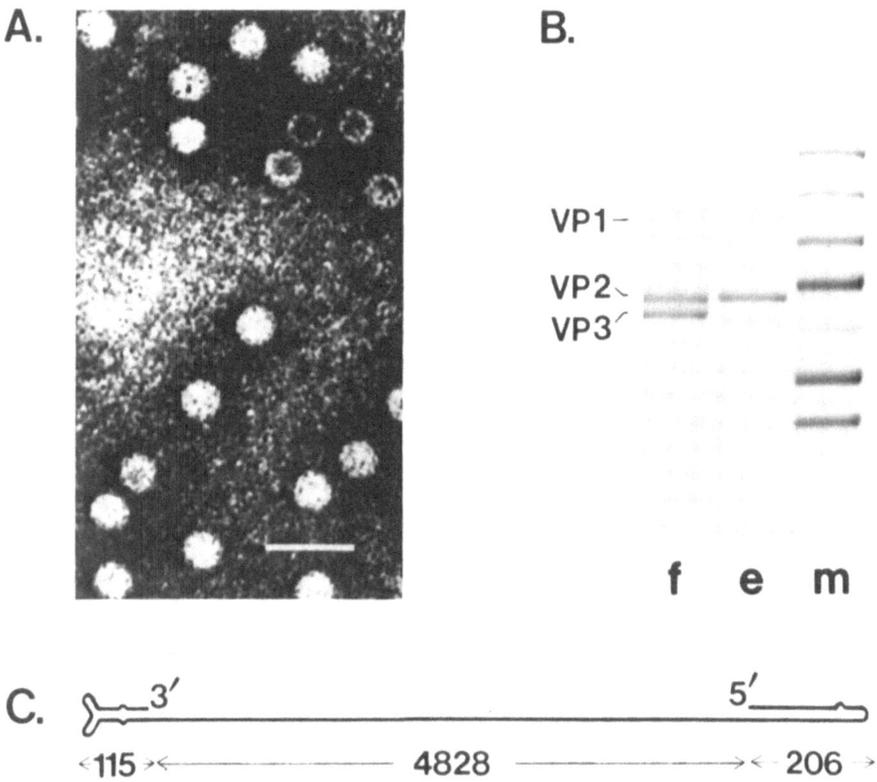

FIGURE 1. Structure of the parvovirus virion. (A) Electron micrograph of purified MVM(p) virions, stained with 3% phosphotungstic acid and photographed at an instrument magnification of 80,000 ×. The horizontal bar represents 50 μm. (B) Proteins of MVM(p) full virions (f) and empty capsids (e) analyzed by SDS-polyacrylamide gel electrophoresis. The viral proteins VP1, VP2, and VP3 are 83, 64, and 62 kdaltons, respectively. Markers (m) are 130, 100, 77, 68, 57, 53, and 49 kdaltons, from top to botton. (Data taken from Tattersall et al., *J. Virol.*, 20, 273, 1976.) (C) Linear representation of the arrangement of duplex and single-stranded DNA sequences in the MVM(p) genome as described in Astell et al.[105] The hairpin structures at each end are drawn 10 × the scale used for the single-stranded portion of the genome.

particle.[29] Both species sediment at 110S, have a molecular weight of 5.5 to 6.2 $\times$ 10^6 daltons[30] and have a particle/infectivity ratio of around 200 to 400 to 1.[31] Although infectious 1.41 g/cm^3 virus is frequently the major form of particle isolated from infected animals, virus stocks grown in tissue culture can contain large numbers of empty protein capsids and defective particles which contain submolar amounts of DNA. The molecular weight of the empty capsids has been estimated at 4.2 $\times$ 10^6 daltons, and they band as sharp peaks at 70S in velocity gradients and at 1.32 g/cm^3 in cesium chloride.[30] Although the relative proportions of full and empty particle can vary depending upon both the cell type and the virus, in vitro empty virions usually outnumber infectious particles by between 2- and 50-fold.[29] Defective particles have a normal capsid component of 4.2 $\times$ 10^6 daltons, but also contain a variable, but subgenomic length of DNA such that their bouyant densities range from 1.33 to 1.39 g/cm^3.[32]

B. Biochemical Structure

The DNA in infectious particles makes up 19 to 37% of the total mass with the capsid proteins probably making up most, if not all, of the residuum. These particles do not appear to contain lipids, carbohydrates, cellular or virally coded enzymes, or low molecular weight histone-type proteins. It is not known if polyamines are present to assist in stabilizing the DNA by charge-neutralization, as has been shown to be the case for members of the *Densovirus* genus of the Parvoviridae,[33] but it seems likely that in many of the autonomous viruses a histone-like function is provided, at least in part, by a specialized, highly basic region at the amino-terminal of the largest of the capsid polypeptides species.[29] Some reports suggest that the viral DNA may be covalently associated with a protein, probably at the 5′ end of the genome, but there is no clear evidence as to the nature or function of such an interaction.[34-36] Classically parvoviruses are regarded as highly stable particles, being resistant to extraction with lipid solvents, pH ranges between 3 and 9, heating to 56°C for 60 min and exposure to the relatively high salt concentrations used in cesium chloride isopycnic gradients.[30] Prolonged storage and repeated freeze-thawing may, however, compromise virus integrity so that, while remaining infectious, some viruses are no longer stable to extremes of heat or high salt concentrations.[37]

Viral characteristics discussed so far appear common to at least most members of the *Parvovirus* genus. However, analysis of the structure of the viral capsid provides a clear indication that the autonomous viruses must be divided into a number of rather disparate groups. Members of the largest group identified to date share the broad features of capsid structure exhibited by the type-species of the genus, Kilham's rat virus (RV). This group includes H-1, RV, MVM, LuIII, PPV, and members of the FPV serotype. Other autonomous viruses such as BPV, LPV, ADV and probably B19 each exhibit a number of characteristics which distinguish them from the RV-like viruses. As will be discussed later, there is evidence to suggest that BPV and LPV may be structurally similar to each other, but the other serotypes appear superficially quite disparate, and it may be that the autonomous parvovirus genus comprises a number of relatively distantly related subgroups. In addition, there are a number of other reported virus isolates, obtained from various animal species such as the goose, chicken, shrimp, horse, and human, which appear to belong to this genus but for which there is little structural information currently available: it is, therefore, impossible to say if these are unique viruses or if they are closely related to any of the previously recognized virus types.

Members of the RV-like group have three major capsid polypeptides in the mature virion (Figure 1B), of which two, VP-1 (83 to 86 kdaltons) and VP-2 (64 to 66 kdaltons) appear to be primary translation products,[38] while the third, VP-3 (60 to 62 kdaltons) is derived by proteolytic cleavage of VP-2.[39,40] Tryptic and chymotryptic peptide analysis has shown that the entire amino acid sequence of VP-2 is present in VP-1.[40] In addition to sequences in common with VP-2, VP-1 contains an amino-terminal region of approximately 17,000 daltons which contains a large number of basic amino acids. This basic region naturally has a profound influence on the isoelectric point of the molecule such that while VP-2 and VP-3 have pI values between 6.6. and 7.2 (for MVM and H-1), the isoelectric forms of VP-1 (MVM) exhibit pI values between 8.2 and 8.5.[41] Although all the viruses in the RV-like group have capsids with a fixed amount of VP-1 (between 12 and 18% of the total capsid protein), the relative proportions of VP-2 and VP-3 can vary dramatically for each virus from preparation to preparation. Highly purified empty particles contain only the primary translation products VP-1 and VP-2, but all purified preparations of infectious virus also contain at least a few copies of the proteolytic cleavage product VP-3. Moreover, if empty virus particles are exposed to the proteases trypsin or chymotrypsin the structural polypeptides remain intact, but if infectious particles are treated with these proteases, VP-1 remains intact while VP-2 can be almost quantitatively cleaved to a VP-3-like peptide.[40] Although these proteases fail to mimic the in vivo cleavage exactly,[29] they do cut in the same part of the molecule, trypsin cutting about 20 amino acids from the N-terminal of VP-2.[42] Thus the presence of DNA in the virion is associated with a change in the conformation or posttranslational modification of VP-2 such that a protease-sensitive site is exposed to the external environment. The biological significance of this change in virion structure, together with the concomitant cleavage of at least a few of the VP-2 molecules, has yet to be evaluated experimentally.

Estimates of the molecular weight of the viral capsid and the weight and stoichiometry of the individual capsid polypeptides predict a particle made up of 60 to 72 protein molecules of which 6 to 9 are VP-1. Since the entire amino acid sequence of VP-2 is contained within that of VP-1, the fact that in full virion preparations the larger molecule cannot be cleaved by proteases, whereas most, if not all, of the smaller species can, suggests that the common body region of these two molecules do not occupy exactly equivalent positions in the mature virion, but that each has its own unique conformation or chemical modification. Although morphologically the capsid appears to be made up of a number of capsomers (Figure 1A), details of its icosahedral structure remain uncertain.

In contrast to those of the RV-like viruses, infectious bovine parvovirus (BPV) and lapine parvovirus (LPV) virions contain four polypeptides. BPV virions purified from infected tissue culture cells have proteins with apparent molecular weights of 80, 72, 62, and 60 kdaltons which make up 6.8, 4.6, 76.8, and 11.8%, respectively, of the total capsid protein,[43] while infectious LPV particles contain polypeptides of 96, 85, 75, and 70 kdaltons, comprising 5, 8, 78, and 9% of the total protein mass, respectively.[44] Interestingly, the capsids of LPV are reported to be 27 to 28 μm, unusually large for a parvovirus,[45] and this coincides with the relatively higher molecular weights of each of the individual LPV capsid proteins. Empty particles of LPV appear to contain relatively little of the 70 kdalton polypeptide (estimated at 2% of the total), while purified full particles contain a rather variable amount of this species (4 to 14% of the total) which can be increased dramatically by in vitro exposure to trypsin (but not to chymotrypsin), suggesting that the smallest structural polypeptide of this virus may also be derived in vivo by proteolytic degradation of a higher molecular weight species. Partial proteolysis of the three higher molecular weight proteins of the BPV capsid with *Staphylococcus aureus* V8 protease revealed extensive amino acid sequence overlap,[43] indicating that, as discussed later for the RV-like viruses, the various capsid proteins are derived from overlapping transcripts encoding the same region of the viral genome. Similarly, the three higher molecular weight capsid proteins of LPV appear to give similar partial V8 maps when analyzed in a two-dimensional system.[44] Although the structural similarities between both the capsid proteins of BPV and LPV and the noncapsid proteins of these viruses, discussed in Section IV, suggest a close phylogenetic relationship, direct comparative studies have yet to be reported. However, these structural features, and the antigenic and DNA-hybridization analyses reported later, do make it clear that BPV is only very distantly related to the RV-like viruses, or to a possible third group of autonomous viruses for which the type species is the recently characterized human virus B19.

When purified from human plasma the B19 virion contains two major capsid proteins, of 83 and 58 kdaltons, of which the latter is by far the predominant species, constituting at least 80% of the total protein mass.[46] Analysis of the DNA sequence of the coat protein genes of B19 confirms that the capsid structure of this virus must also be substantially different from that of the RV-like viruses. However, prokaryotic expression studies have shown that a region of continuous open reading frame in the B19 genome encoding some 284 amino acids, contains antigenic determinants present on both the 83 and 58 kdalton polypeptide. Thus, as with the other viruses in this genus, the B19 capsid proteins appear to share overlapping amino acid sequences.[46]

Aleutian disease virus (ADV) exhibits a fourth type of capsid structure. ADV-Gorman and Utah-1 strains of ADV can be grown in vitro in Crandell feline kidney cells with variable efficiency. When propagated in vitro, the capsids of these viruses have two polypetides of 85 and 75 kdaltons which share overlapping peptide composition and of which the 75 kdalton species is slightly the more abundant.[47] However, when in vitro grown ADV-Gorman particles were first exposed to trypsin and the still intact full particles then reisolated by centrifugation on cesium chloride gradients prior to sodium dodecylsulfate polyacrylamide gel analysis (SDS-PAGE), the 85 and 75 kdalton proteins had been degraded *in situ* to a number of much lower molecular weight protein species.[48] Similarly, when virions from Utah 1, Pullman, and DK strains of ADV were purified from infected mink tissues, the only structural polypeptides which could be identified had molecular weights between 27 and 30 kdaltons. Although the in vitro trypsin-treated virions have a somewhat reduced infectivity, sera from infected mink contain high titers of antibodies directed specifically against the proteolyzed form of the virus, suggesting that in vivo this type of degradation is common.[48] Such extreme sensitivity of the structural proteins to trypsin is clearly unique in the autonomous parvoviruses, but lack of sequence information, antigenic comparisons, or DNA-hybridization studies makes it impossible to assess how closely ADV is related to other members of the group.

C. Antigenic Structure

Classically, parvoviruses were identified and differentiated from each other by the specificity of the antigenic determinants exhibited on their intact capsids. Since most of these viruses are known to agglutinate erythrocytes from one animal species or another, serotypes are conveniently distinguished by antibody-mediated inhibition of hemagglutination (HI). Although disparities can occur, overall results from serum neutralization studies usually mimic HI data, probably because the multiple loci which determine these parameters are either overlapping or at least closely juxtaposed on the virion surface. On the basis of such tests the parvoviruses have been separated into a number of distinct serotypes, summarized in Table 1, which share practically no HI or neutralizing antigens. However, if used with heterospecific antisera, these tests are rather insensitive to minor antigenic drift between virus isolates or host range variants in a particular antigenic group, since they average the influence of antibodies directed against many different determinants. In some serotypes, most notably the FPV-group, panels of monoclonal antibodies have been used to overcome these limitations and permit the fine dissection of antigenic drift in a changing virus population. Using 13 monoclonal antibodies raised against CPV and 8 monoclonal antibodies against FPV, Parrish and Carmichael[49] were able to demonstrate clear differences between the four viruses in this serotype (FPV, CPV, MEV, and RPV), and between many individual isolates of the same subtype. However, of the total 21 neutralizing antibodies used in this study 14 reacted with most isolates, reaffirming the close overall antigenic relationship of these host range variants to each other. When fourth and fifth generation cell culture-passaged stocks of CPV and FPV were added to cells in the presence of various individual, homologous monoclonal antibodies, preexisting virus variants could usually be selected at frequencies between 10^{-4} and 10^{-7}. In all cases these variants now failed to react with the selecting antibody, but reacted to different degrees with other monoclonals. Studies of this type together with competitive binding assays between the various antibodies allowed these authors to cluster the monoclonals into groups which influenced the binding of other members of the same group to a greater or lesser extent, but which did not influence binding of antibodies from another group. In this way they showed that the capsid surface of the FPV-serotype viruses contains several different, but overlapping, neutralizing antigenic sites, each comprising a number of different, but overlapping, neutralizing epitopes.

All 21 of the murine monoclonal antibodies obtained in the Parrish and Carmichael study were capable of plaque neutralization.[49] Since there was no obvious pressure applied to ensure the selection of such antibodies, this suggests that a very high proportion of the antigenic determinants accessible on the surface of the virus can influence infectivity, and thus that the surface structure of one virus serotype is likely to be quite different from that of another. However, nonneutralizing surface epitopes do exist in canine parvovirus (CPV), as demonstrated by Burtonboy et al.[50] Of the 40 anti-CPV rat monoclonal antibodies obtained by these authors, four did not neutralize viral infectivity, and these same four failed to influence binding of any of the neutralizing antibodies.

III. STRUCTURE OF THE VIRAL GENOME

A. DNA Structure and Sequence

Each autonomous parvovirus virion contains a single copy of a linear, nonpermuted DNA molecule about 5 kbases in length in which a long single-stranded coding region comprising over 90% of the genome, is bracketed by shorter terminal palindromic regions, capable of folding into hairpin duplexes.[51] The packaged strand can be predominantly of one polarity — invariably the complement of the coding sense — or a mixture of strands of both polarities packaged in separate virions.[1] By convention the genome is always drawn with the coding sense (mRNA 5′ to 3′ direction) going from left to right. For those viruses packaging predominantly a single-sense DNA strand, this places the 3′ end of virion DNA on the left and its 5′ end on the right as depicted in Figure 1C. This convention will be used throughout this article, to assign ends to monomeric units of both single- and double-stranded forms of the viral genome.

Several autonomous parvovirus DNA molecules have been molecularly cloned and partially or completely sequenced. The viruses for which complete DNA sequences are available are MVM(p), 5149 nucleotides,[52,53] MVM(i), 5085 nucleotides,[53,54] and H-1, 5176 nucleotides.[55] Detailed sequence information about the viral termini of B19 is still lacking, but the sequence of the entire coding region is now known, and this genome is at least 5300 and probably closer to 5500 nucleotides long.[56] Sequence information presented recently for BPV suggests that its genome is also about 5500 nucleotides long.[57] Size estimates for other viruses based predominantly on the electrophoretic mobility of the denatured viral strands or restriction endonuclease fragments of the replicative forms suggest that few, if any, viruses have genomes smaller than MVM(i) or larger than B19.[58] This careful conservation of genome length suggests that there are fairly rigid constraints on the maximum length of DNA which can be encapsidated, although we know of no experimental data which directly addresses this problem.

Although the genomes of all autonomous parvoviruses share certain common characteristics diagnostic of the virus family, a somewhat superficial analysis of capsid structure has already suggested that the genus may contain a number of disparate virus groups, and analysis of the viral genome lends additional support to this supposition. In the absence of full DNA sequence information, DNA hybridization and heteroduplex mapping studies can provide some insight into the genetic relationship of one virus serotype to another, but little of this work has been fully quantitated and much of it remains unpublished. Of the studies available, one by Banerjee et al.[59] used heteroduplex mapping of RF DNA from MVM, RV, H-1, and LuIII to show that complementary strands from these viruses were capable of reannealing along a continuous stretch covering approximately 70% of their genome length, predominantly involving the left-hand end of viral DNA. In contrast, the genome of BPV completely failed to hybridize to these viruses or to AAV under the same annealing conditions.[59] Southern blot hybridization data from our laboratory essentially confirms these observations and also shows that CPV and porcine parvovirus (PPV) share limited homology with the rodent group. This, once again, allows us to cluster the rodent viruses with LuIII, CPV, and PPV into a single group of rather obviously related viruses (designated the RV-like viruses), but leaves BPV by itself. The RV-like viruses also show trace homology when hybridized with nick-translated probes derived from the B19 genome, although in this case the cross hybridization is extremely weak, rarely exceeding 0.1% of the homologous reaction.[58] Interestingly, in the same series of experiments, these B19 probes failed to hybridize to BPV or AAV-2 genomes even at a stringency of Tm-35°C. Clearly B19 is only very distantly related to any of the Parvoviridae examined so far, although recent evidence suggests that there may be an antigenically distinct human enteric virus to which it cross hybridizes rather efficiently.[60]

Both the 3′ and 5′ termini of all parvovirus genomes analyzed to date contain palindromic nucleotide sequences which can exist in the form of relatively stable hairpin duplexes in the single-stranded form of viral DNA. In the rodent viruses MVM, KRV, H-1, and H3, the 3′ hairpin structure of virion DNA comprises 115 or 116 nucleotides,[61] while the 5′ hairpins of MVM and H-1 contain 207 and 242 nucleotides, respectively.[52,55,62] Unlike the adeno-associated viruses, this group of autonomously replicating rodent viruses do not possess inverted terminal repeats, but have a unique primary DNA sequence at each terminus. Initiation and termination of DNA replication occurs in the terminal regions, as discussed later, and the efficient replication and encapsidation of Type I defective genomes of MVM, which have very large internal deletions, demonstrates that all critical *cis*-acting sites necessary for these processes are located entirely within 200 to 300 nucleotides of each genomic terminus.[32] The 3′ nucleotide sequences of MVM(p), H-1, KRV, and H3 have been compared by Astell et al.[61] Although there are some minor differences, the sequence of the first 150 nucleotides of each virus is essentially the same, and, since sequence homology at this level is not maintained throughout the viral DNA, this suggests a strong pressure to conserve this particular region. A total of 104 of the first 115 nucleotides of MVM can be based-paired to form a stable Y-shaped hairpin, and it may well be that it is the constraint implied in maintaining this overall conformation which resist genetic drift, rather than absolute linear sequence requirements. A bubble in the duplex stem created by a mismatch between nucleotides 23 to 26 and 88 to 91 is a common feature in all four viruses, suggesting that it may also be an essential element of the terminal structure.

Restriction mapping and DNA sequence studies showed that the right-hand termini of MVM and H-1 RF DNA can exist in two alternative sequence orientations, termed "flip" and "flop", which occur with equal frequency in the monomeric, double-stranded, replicative form (RF) of the genome isolated from infected cells.[52,62] The 5′ end of single-stranded MVM viral DNA has also been shown to exist in both orientations.[63] These two orientations are only apparent because the terminal sequences have small asymmetric loops near the axis of symmetry of what are otherwise perfect palindromes. Sequence inversions of this type are also found for both terminal repetitions of packaged AAV DNA, but are not found at the left-hand ends of either RF or virion DNA from MVM, although this terminus also contains asymmetries. The presence or absence of such inversions is of primary importance when considering possible models of DNA replication, as will be discussed below.

The RF form of MVM(p) DNA has at least an additional 18 nucleotides at its right-hand terminus which are not present on the 5′ end of the DNA strand encapsidated into virions.[52] The additional nucleotides are the complement of a sequence in the viral genome (residues 4923 and 4940) that lies immediately to the 3′ side of the 5′ hairpin structure, and its seems likely that, rather than nicking at the normal cutting site for virion DNA (nucleotide 5149) during the formation of RF DNA the 5′ nick occurs on the complementary strand 18 bases from nucleotide 5149 at nucleotide 5167. This observation is most easily explained by suggesting that for MVM the potential site-specific nickases involved in the replication of the 5′ ends of RF and virion DNA are different, as will be discussed later.

In contrast to the dissimilar sequences found at each end of the RV-like viruses, the partial sequence information available from molecularly cloned, albeit deleted, forms of the B19 termini[56] suggest that these are extensively homologous and may even be slightly imperfect, inverted terminal repeats. They also appear to be very much longer than those of the rodent viruses, with the left-hand palindrome comprising at least 178 nucleotides and that at the right-hand end at least 240 nucleotides. We have recently cloned a form of the right-hand terminus of B19 which is at least 70 to 80 bases longer than this, making the probable length

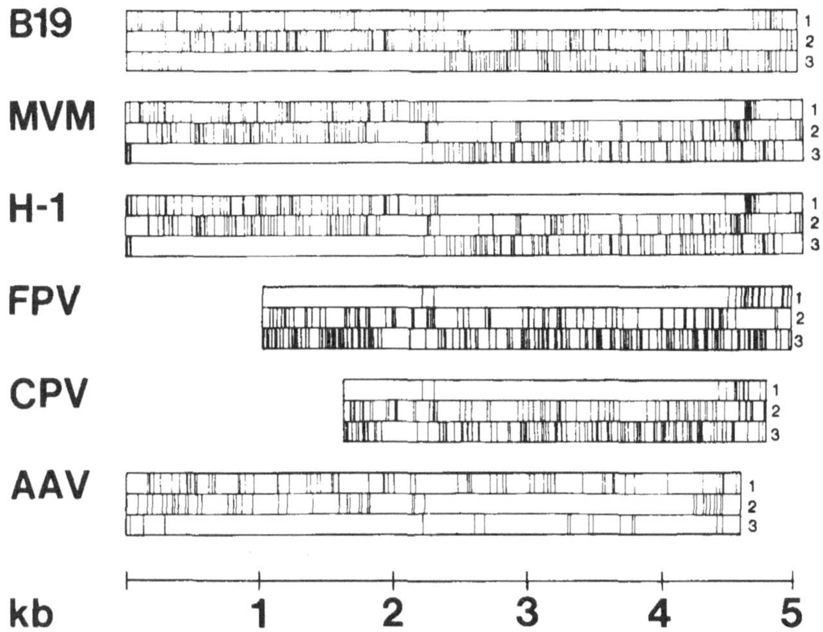

FIGURE 2. Comparison of known DNA sequences of members of the Parvoviridae. The sequence of the coding strand of each virus, as far as currently available, is shown as a series of vertical bars representing predicted translational termination codons in each of the three reading frames. (Adapted from Shade et al., *J. Virol.*, 58, 927, 1986.)

of the hairpin sequence around 320 nucleotides. Unfortunately, using plasmid DNA to obtain sequence data for viral termini is problematic since the central regions of many palindromes become deleted, and plasmids carrying such deletions appear to be preferentially replicated by the host bacteria.[64,65]

B. Organization of the Genome

In all parvoviruses analyzed to date, all of the protein coding regions appear to be clustered on one of the DNA strands which is, by definition, the plus strand. In the case of parvoviruses which encapsidate strands of one sense, this coding strand is the complement of that virion DNA molecule. Where transcription data is available, it confirms the transcriptional use of only one strand. The blocks of open reading frame available in the plus strand of the sequenced virus genomes are diagrammed in Figure 2. In each case there are two large open reading frames (ORFs) which together span almost the entire genome, and a number of smaller ORFs, the exact size and location of which varies somewhat from virus to virus. As will be discussed further in the next section, in each case the long left-hand ORF is known to encode at least one major nonstructural protein, while the right-hand ORF provides most of the sequence expressed in the various capsid polypeptides. Comparison of the published sequences of MVM(p) and MVM(i) show that there are 163[54] or 175[53] nucleotide differences between these two strains, depending on the exact source of MVM(i). Figure 3 illustrates the influence of these nucleotide changes on the amino acid sequences expressed in the two major ORFs. Most of the nucleotide differences in these regions are third-base changes in ORF codons, and so do not influence protein sequence. However, the distribution of these changes is nonrandom, and very few are found between nucleotides 2000 and 3000. This suggests that there are additional constraints operating in this region which deter even third-base changes, and one obvious way to account for this is to suggest that the small ORFs located in frame 2 also encode protein. This has been shown to be the case for the region

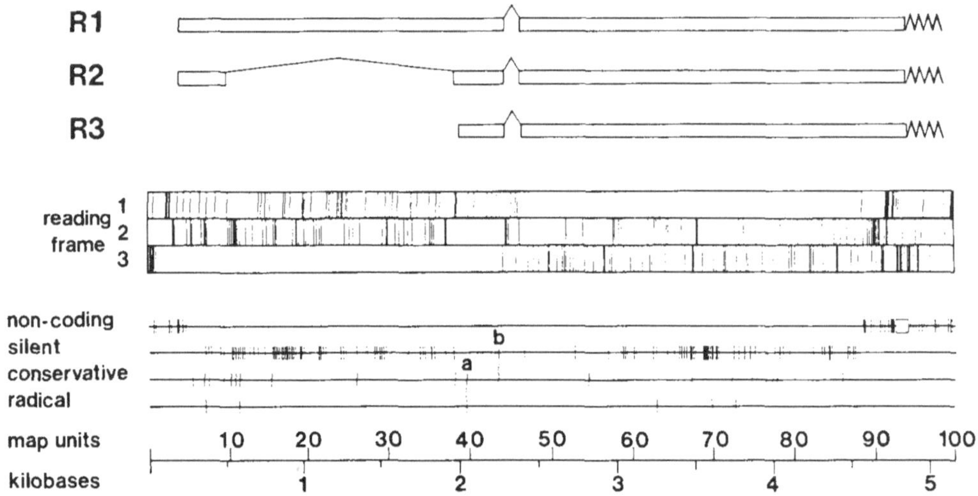

FIGURE 3. Comparison of the genomes of MVM(i) and MVM(p). Genetic map of MVM showing the distribution of nucleotide differences between the allotropic variants MVM(p) and MVM(i). The viral genome is displayed with the 3′ end of virion single-stranded DNA, the negative strand with respect to transcription, on the left at zero map units. At the top of the diagram the three major cytoplasmic transcripts R1, R2, and R3 are represented by double lines with single, wavy lines indicating their polyadenylated tails, and single, straight lines indicating the introns spliced out in the production of mature message. Beneath the transcripts a block diagram in which vertical bars represent translation termination codons, shows the regions of open reading frame in all three frames of the transcribed (or complementary) DNA strand. A horizontal line labeled "noncoding" uses vertical bars to indicate nucleotide differences between MVM(p) and MVM(i) in the noncoding region of the genome while the three lines below it similarly indicate changes in the coding sequences. Nucleotide differences which do not change the encoded amino acid (designated "silent") are depicted separately from those which cause "conservative" or "radical" changes in the specified amino acid. Bars labeled "a" and "b" indicate the nature of the amino acid changes in the two positions where the viral genome is known to encode proteins in both the available open reading frames. "a" denotes a radical change in the sequence of NS-2 coincident with a conservative change in NS-1, and "b" denotes a conservative change in the sequence of NS-1 which does not effect the sequence of NS-2 (see Figure 5 for further explanation). (Adapted from Astell et al., *J. Virol.*, 57, 656, 1986.)

between map units 38 to 44 in reading frame 2 of MVM and H-1,[66] but proteins encoded by the other small ORFs have yet to be identified. CPV, FPV, and B19 lack a small ORF homologous to that known to encode protein in MVM and H-1 (frame 2, map units 38-44), but conservation of nucleotide sequence between CPV and FPV in various other small ORFs strongly suggest that some of these minor open frames may well be utilized to encode protein.

In eukaryotes, RNA polymerase II transcription control regions are usually characterized by a TATA box which occurs approximately 30 nucleotides upstream of the actual site of initiation of the ribonucleic acid (RNA) chains. While the precise function of this sequence is not clear, it is generally thought to specify the start site for transcription. The plus-strand sequences of MVM and H-1 contain several of these signals, two of which are located upstream of the known transcription start sites at map units 4 and 39, while the published sequence for CPV also contains a TATAAA box at nucleotide 318, in an equivalent position to the TATAAA boxes at map unit 39 in MVM and H-1. In FPV this sequence has been mutated to TGTAAAT (a less usual, but not unprecedented transcription start site), and there is a TAAAATA sequence not found in MVM, H-1, or CPV located 15 bases upstream of the TGTA site. A major FPV transcript has been shown to initiate just downstream of these signals and its seems probable that one or other of them constitutes part of the major promoter region. A second eukaryotic consensus, GGPyCAATCT (the CAAT box) characteristically located around 30 nucleotides upstream of the TATA box, has also been

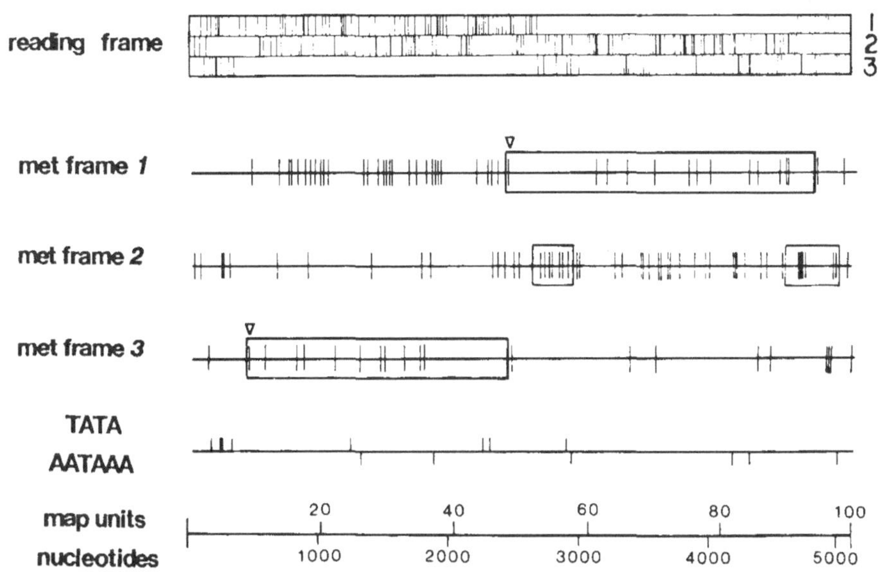

FIGURE 4. Coding strategy of the B19-Au genome. The bar diagram represents translation termination signals in all three reading frames of what is presumed to be the plus strand of the B19 genome. Lines labeled "Met frames 1 to 3" use vertical bars to indicate the positions of methionine codons in the three reading frames, and boxed segments represent the major regions of open reading frame in this strand. Positions of TATA sequences and polyadenylation signals (AATAAA) are aligned for comparison with these blocks of coding sequence.

implicated in the efficient initiation of RNA polymerase II transcripts in vivo. The known H-1, MVM, CPV, and FPV promoters lack this sequence in the appropriate position, although the promoters at map unit 39 in MVM and H-1 have a CCAAT sequence 87 nucleotides upstream of the TATAAA box. However, in CPV and FPV the sequence in the equivalent position is mutated to TGAAT, suggesting that such signals are not essential for efficient transcription in these viruses, and emphasizing the rather unusual nature of the viral control regions. There is now good evidence to suggest that the virally coded NS-1 protein in some way regulates the rate of transcription from the promoter at map unit 39 in H-1,[67] and may also be able to influence the rate of transcription from its own promoter at map unit 4.[68] Whether these interactions are direct or indirect remains to be seen, but preliminary deletion-mutant studies and comparative sequence analysis suggest that a region extending approximately 116 nucleotides upstream of the TATA box at map unit 39 may be involved.[68] All cytoplasmic viral mRNA molecules terminate in variable length poly-A sequences which are not directly encoded in the genome. However, polyadenylation signals (AATAAA), located downstream of coding regions direct cellular machinery to append the polyadenine sequence. Such signals in MVM, H-1, and probably CPV and FPV, are restricted to the extreme right-hand end of the genome.

In contrast to the relatively simple situation in the RV-like viruses, a search of the plus strand of B19 reveals the multiple potential transcription start and polyadenylation signals illustrated in Figure 4. Five TATA sequences are clustered at the extreme left-hand end of the viral genome, in the position of the single sequence seen in the rodent viruses, and there are additional TATA boxes at nucleotides 1225, 2247, 2308, and 2986. Interestingly, three of these have appropriately spaced upstream CAAT sequences, but at present there is no direct information as to which of these actually function as part of transcriptional promoter sequences. Six potential polyadenylation signals are found at nucleotides 1303, 1872, 2935, 4168, 4307, and 4990, but again, how many of these are used remains to be established.

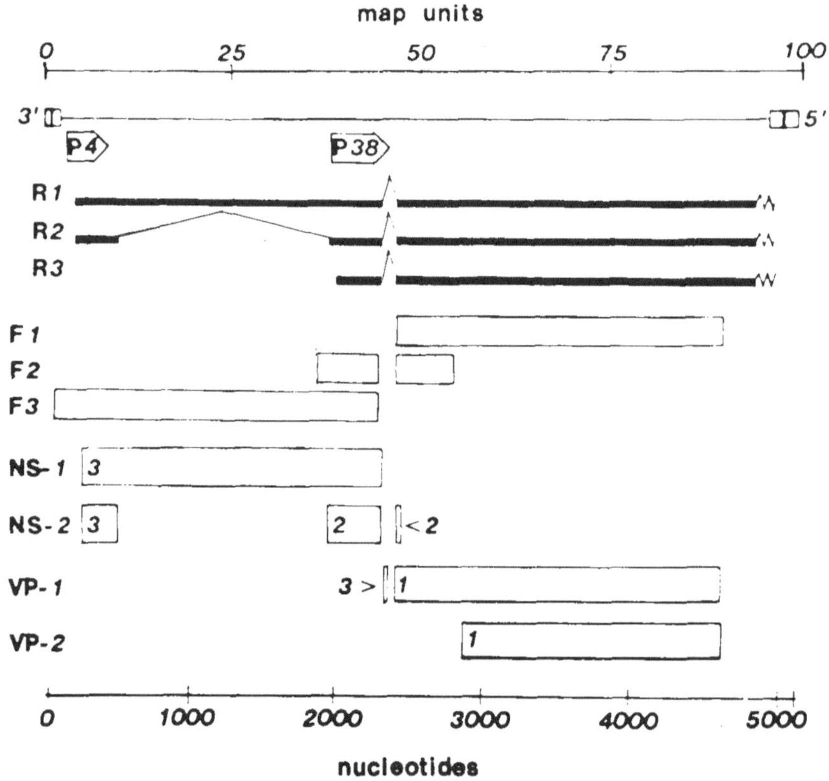

A

FIGURE 5. Coding strategy of the MVM genome. (A) The cytoplasmic transcripts of MVM, denoted R1, R2, and R3, are aligned beneath a line diagram of the viral DNA strand which illustrates the extent of the 3′ and 5′ terminal hairpin palindromes and the positions of the two promotors at map units 4 and 38. Major blocks of open reading frame in the transcribed DNA strand are depicted for each of the three possible reading frames (F1, F2, and F3) and the sequences encoded in the viral proteins NS-1, NS-2, VP-1, and VP-2 are illustrated with numerals to designate which reading frame is expressed in each part of the molecule. (B) A block diagram of the major open reading frames in all three translational phases of the transcribed DNA strand is marked A, B, C, and D to indicate the sequences and reading frames of the fragments expressed as fusion proteins in bacteria.[66] Proteins immunoprecipitated from the total translation products (T) of mRNA from uninfected (u), and MVM(p)-infected (i), A9 cells with antibodies against the protein fragments A through D, are compared with those precipitated using antibodies directed against the capsid proteins (encoded by the region marked E).

Preliminary sequence information for BPV also reveals multiple potential transcriptional promoters and polyadenylation signals scattered throughout the viral genome.[57]

IV. CODING STRATEGY OF THE VIRAL GENOME

A. Viral Gene Expression

The most complete analysis of viral transcription has been carried out in MVM(p)[69] although the picture which is emerging may well be common to most, if not all, the RV-like viruses. MVM(p) encodes two overlapping transcription units with separate promoters near the left end (map unit 4) and middle (map unit 38) of the viral genome. As seen in Figures 3 and 5, three major spliced and polyadenylated RNAs have been identified, 4.8, 3.3, and 3.0 kbases in length (designated R1 to R3) which constitute approximately 10 to 15, 15 to 20, and 65 to 70%, respectively, of the total MVM specific RNA as analyzed by

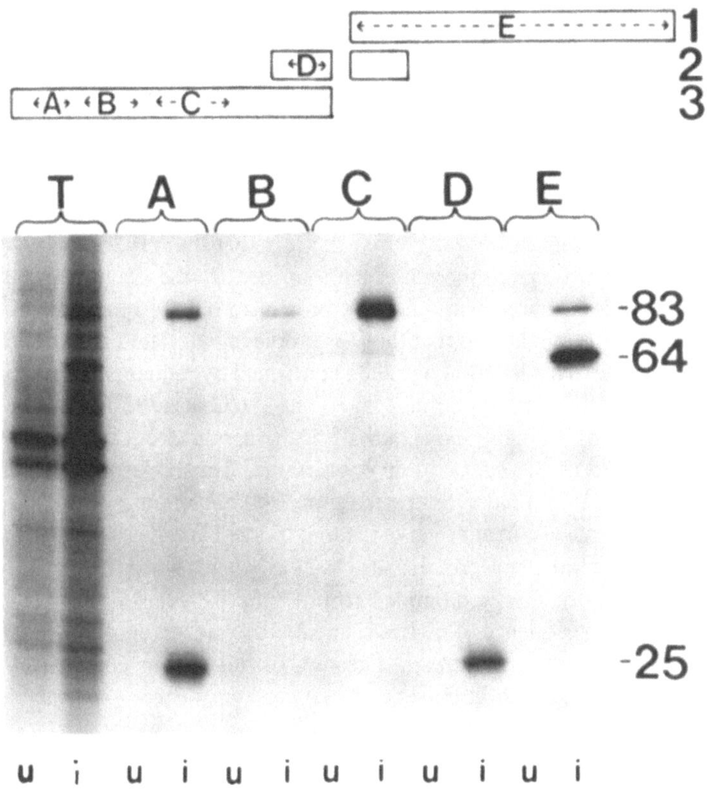

FIGURE 5B

quantitation of Northern blots.[69] All three species are transcribed from the virion (-) strand of DNA and all have a short intron sequence between 46 to 48 map units removed. Transcripts R1 and R2 originate from the promoter at map unit 4, and have their 5′ ends at nucleotide 201 ± 5, while the most abundant viral RNA molecules (R3) start at nucleotide 2005 ± 5[70] and are the products of the promoter at map unit 38. Although transcription data is not yet available for B19 or BPV, the multiple TATA boxes and polyadenylation signals distributed throughout these genomes suggest that they may give rise to a rather different series of RNAs.

In vitro translation of mRNA from cells infected with MVM, H-1, or LuIII gives four major virally coded proteins.[66] In MVM two of these co-migrate on SDS-polyacrylamide gels with the viral capsid proteins VP-1 and VP-2 (83 and 64 kdaltons, respectively), and are indistinguishable from these proteins purified from viral particles both antigenically and by peptide map analysis.[38] Synthesized in vitro in approximately the same 1:5 ratio as the VP-1 and VP-2 species found in assembled empty capsids, these molecules appear to be the primary translation products of the viral structural proteins. The other two proteins, designated NS-1 (83 kdaltons) and NS-2 (25 kdaltons), are not related to the capsid proteins, but are recognized by sera from animals infected with a number of different autonomous parvovirus serotypes. In Figure 5A the map coordinates of the three major transcripts of MVM are aligned with the viral genome, and with the major blocks of open reading frame in the plus strand.

In vitro translation of BPV-infected cell RNA, however, yields at least three structural polypeptides which co-migrate with the capsid proteins VP1-3 (80, 72, and 62 kdaltons), and a major nonstructural protein, termed NP-1, of 25 kdaltons.[43] Additional virally coded proteins are predicted from the DNA sequence,[57] but have yet to be identified in vitro or in vivo.

B. The Structural Polypeptide Gene

As mentioned previously, the various capsid proteins of all the autonomous parvoviruses analyzed to data are made up from overlapping amino acid sequences, such that the larger molecules contain the entire sequence of all smaller capsid species, but have unique amino-terminal extensions.[40,73] Paradiso et al.[42] used the fact that the VP-2 to VP-3 cleavage can be mimicked in vitro with trypsin to enable them to identify the amino-terminal peptide of the H-1 VP-2 molecule. Although this peptide proved to have a blocked amino-terminus, its amino acid composition indicated that the VP-2 molecule initiated at the AUG at position 2797 in the H-1 genome. This location was then confirmed by sequencing the amino-terminal region of VP-3 created by this tryptic cleavage. The AUG at 2797 is the first such codon in the major open reading frame in the right half of the genome, but it is approximately 400 nucleotides downstream of the start of this open frame. Thus ribosomes loading on to the 5′ end of R3 transcripts might have to traverse some 790 nucleotides of mRNA, and to read through three upstream AUGs (in alternative reading frames) in order to initiate the translation of the most abundant viral gene product at 2797. However, these three AUGs are now known to be spliced out of the predominant form of the R3 transcript (which presumably encodes VP-2), leaving a sequence of 694 ± 5 nucleotides upstream of the initiating AUG.[72]

At least in the RV-like viruses, trypsin digestion of full virions cuts the VP-2 molecule at two sites, releasing two small peptides and leaving a sequence which is extraordinarily rich in glycine at the amino-terminal of VP-3. The accessibility of these two cleavage sites to trypsin suggests that these sequences are located on or close to the outside surface of the virion. In H-1 and MVM, glycine occurs at 13 positions in the first 24 residues of trypsin-derived VP-3 molecules, and in CPV and FPV at 14 of the first 24 residues. Preliminary DNA sequence data for BPV has also identified a similar sequence in what is presumed to be the capsid gene.[57] This region, which exhibits 68% exact homology with the equivalent sequence in H-1, contains 12 glycines in a stretch of 24 amino acids. Since few sequences in BPV appear to share this level of homology with the RV-like viruses, the conservation of this peptide strongly suggests that, in BPV, a post-assembly cleavage of one of the virion proteins is likely to occur. The capsid genes of B19 and the adeno-associated viruses do not, however, contain an exactly equivalent glycine-rich sequence, and there is no evidence for post-assembly cleavage of virion proteins.

The entire coding sequence of the MVM capsid proteins VP-1 and VP-2 can be translated from R3 transcripts.[71] Since both of these are primary translation products, there must be a mechanism by which the translation apparatus of the cell can select two different AUGs. Recently, Jongeneel et al. showed that in many cDNA clones derived from MVM(i) transcripts, nucleotides* 2280 and 2376 are juxtaposed.[72] As seen in Figure 6A, 5′ and 3′ consensus splice sequences occur at the positions necessary to mediate such a translocation, and R3 transcripts carrying this sequence in MVM would contain only a single, apparently unused, AUG at position 2504 (absent from the H-1, CPV, or FPV sequence) upstream of the VP-2 initiation codon at 2794. These R3 transcripts presumably encode the major capsid protein VP-2. However, in order to synthesize VP-1, an initiation codon substantially upstream of this site must be used, and cDNA sequencing data from two laboratories[74,75] indicate that this is achieved by using a different arrangement of splice sites. Thus transcripts encoding VP-1 apparently ignore the 5′ splice site used for VP-2 and so encounter an additional AUG triplet at nucleotide 2286 which falls in the intron of VP-2. Protein synthesis presumably initiates at this AUG and allows translation of a decapeptide (MAPPAKRAKR)

* MVM(p) nucleotide numbers used throughout, unless otherwise indicated.[53]

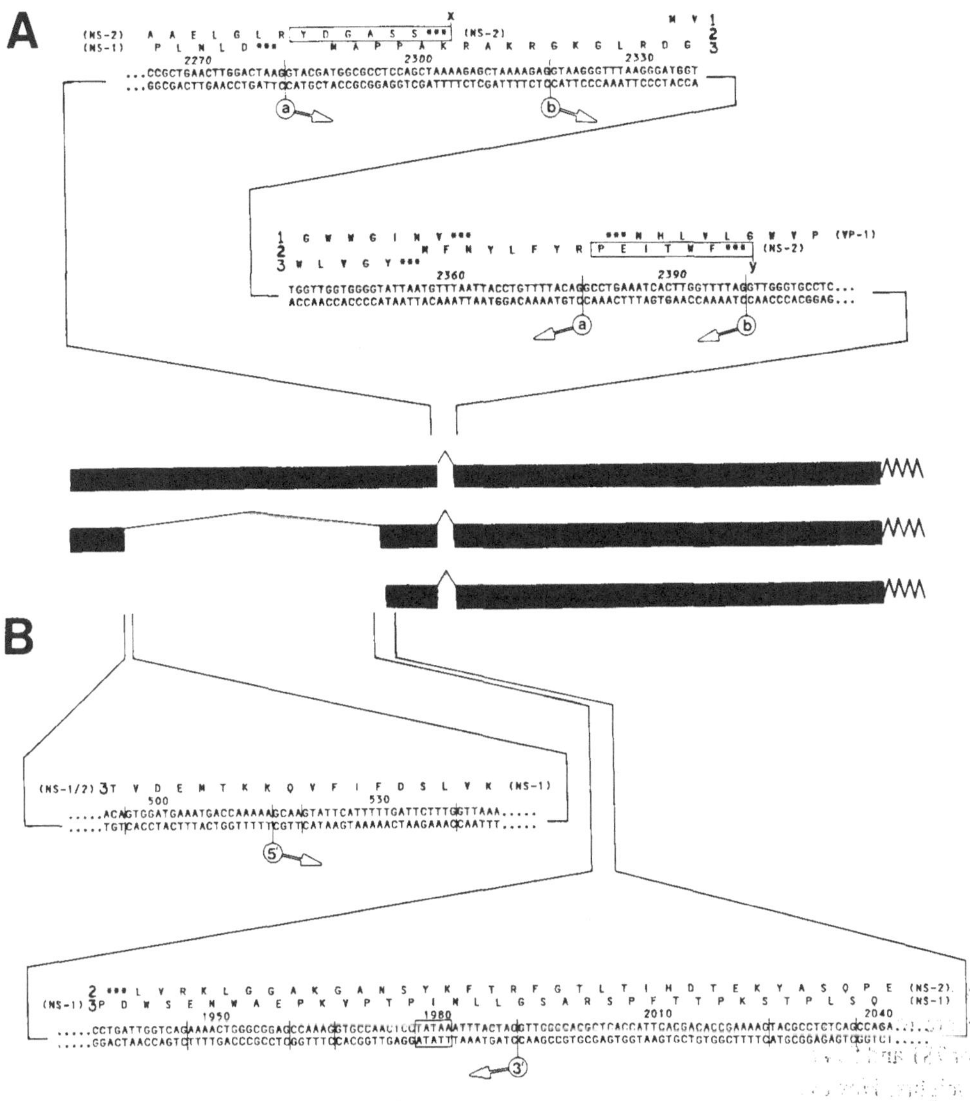

FIGURE 6. Anatomy of the MVM splice junctions. (A) A small splice is present in all of the cytoplasmic transcripts of the virus. In approximately 80% of the transcripts the 5′ donor sequence at position 2281 (immediately downstream of the translation termination codon for NS-1 in frame 3) is spliced to the 3′ acceptor at 2377 (both designated "a"), while in the remaining 20% of transcripts the 5′ and 3′ splice sites marked "b" (at nucleotides ,nts, 2317, and 2399, respectively) are used. Differential expression of the capsid proteins VP-1 and VP-2 is probably controlled by this alternative splicing strategy. VP-1, encoded by transcripts with a splice of the less common "b" type initiates at the ATG in frame 3 at position 2286 and expresses ten amino acids from this region before splicing into frame 1 at the beginning of a major open reading frame which occupies most of the right half of the viral genome. VP-2, encoded by the more abundant transcripts with splices of type "a", does not initiate until well downstream of the coding region detailed here, since the splice removes all of the possible ATG codons. R2 transcripts which contain splices of type "b" terminate translation upstream of the 5′ splice site, but transcripts with splices of type "a" program an NS-2 molecule in which the sequences expressed in frame 2 on either side of the splice are joined. Translation then terminates with an amber codon at position 2396, but if this single codon were suppressed an additional 96 amino acids could be expressed in this frame, terminating at a TAA codon at position 2686. (B) The position of the 3′ and 5′ splice junctions determining the major intron of the R2 transcript are indicated, together with the genomic nucleotide sequence at these junctions. An unusual 5′ splice site (at nt. 515), in which the bases GC substitute for the more usual GT sequence at the intron side of the site, is used despite the presence nearby (at nts. 495, 519, and 540) of a number of other potential 5′ donor sequences which fit more closely to the various proposed consensus sequences. The 3′ splice site used by R2 (at nt. 1990) transfer the coding sequence from frame 3 into frame 2. Other consensus 3′ sites in this vicinity (at nts 1945, 1959, 1965, 2024, and 2036) could transfer the coding sequence into alternative reading frames, but these sequences are ignored. The R1 transcript is not spliced in this region of the genome.

encoded in frame 3 before the RNA splices to a new reading frame using alternative 5′ and 3′ splice sites at nucleotides 2316 and 2398, respectively. This allows the decapeptide in frame 3 to be spliced into frame 1 near the beginning of the major open reading frame known to encode the bulk of the capsid proteins (see Figure 5). This splice removes 82 nucleotides from the genomic sequence and thus gives rise to an mRNA 14 nucleotides longer than the major R3 transcript encoding VP-2. Assuming VP-1 initiates at nucleotide 2286 and terminates in frame 1 at nucleotide 4555, it will have a coding sequence of 2187 nucleotides and contain 729 amino acids, while VP-2 molecules initiating at position 2794 and coterminating with the VP-1 proteins at 4555 will express 1761 nucleotides, encoding a protein of 587 amino acids. Both of these sequences are thus reasonably compatible with the apparent molecular weights of VP-1 and VP-2 (83 and 64 kdaltons) determined by SDS-polyacrylamide gel electrophoresis. Moreover, according to this scheme, the VP-1 specific sequence, which in MVM is responsible for a difference of 19 kdaltons between VP-1 and VP-2 and for a change in pI from between 7.0 and 7.2 to between 8.2 and 8.5,[41] comprises 142 amino acids of which 15 are lysine (10.6 mol %) and 9 arginine (6.3 mol %).

Differential splicing mechanisms similar to the one just described for MVM are thought to be employed by all the RV-like viruses to encode capsid proteins of different sizes from a single gene sequence. While we do not know how this feat is accomplished for other viruses in the genus, in AAV-2 there is evidence to suggest that regulation can also occur at the translational level, with one of the less-abundant capsid proteins (B) being synthesized from an unusual ACG initiation codon present in the same, spliced 2.3 kbase transcript upstream of the ATG codon used to initiate the major capsid protein (C).[76]

In B19 the organization and expression of the capsid genes is likely to be somewhat different to that of MVM or AAV. Two major structural proteins of 83 kdaltons (VP-1) and 58 kdaltons (VP-2) have been described, and prokaryotic expression studies confirm that both of these peptides map to the large open reading frame in the right half of the viral genome between nucleotides 2441 and 4787 and share overlapping sequence.[46] At present no direct transcription data is available since there is no adequate in vitro growth system for this virus. As shown in Figure 4 there is AUG codon located at the beginning of the right-hand ORF at nucleotide 2444. The next in-frame AUG is at position 3125, and this whole stretch of potential coding sequence terminates at nucleotide 4787. If the B19 VP-1 were to initiate at the first position and its VP-2 at the second, this would give molecules of 781 and 554 amino acids, which is clearly compatible with the observed apparent molecular weights. However, there are three other AUGs upstream of nucleotide 2444 but downstream of the TATA box at 2308 which is presumed to be part of the promoter for this transcription unit. Thus, by analogy with the RV-like viruses, there may well prove to be a small splice in this region, removing some of the redundant AUGs, or perhaps even juxtaposing a short sequence from an alternative open reading frame upstream of the major ORF as the amino-terminal of VP-1. In a VP-2 transcript originating from the same promoter and initiating at nucleotide 3125, the problem of multiple AUG codons would be much more dramatic than in MVM, since the RNA would contain 16 out-of-frame potential initiator codons between the 5′ end of the mRNA and the proposed initiation site. It is therefore of particular interest that B19 appears to be an additional promoter in this region, with a CAAT box at 2950 and a TATA box at 2986. The codon at nucleotides 3125 is the first AUG downstream of this promoter, and it would thus seem from simple perusal of the sequence that in B19 the mRNAs for VP-1 and VP-2 molecules may possibly be transcribed from different promoters.

Comparison of the DNA sequence encoding the capsid genes in B19, MVM and AAV-2 reveals a conserved sequence in the VP-1 specific region of these molecules. Over a stretch of 30 amino acids, encoded by nucleotides 2407 to 2496 in frame 1 for MVM(p), 2815 to 2905 in frame 1 for B19, and 2341 to 2431 in frame 1 for AAV-2, these viruses share homologies, at the protein level, of 47% (between B19 and AVV), 50% (between B19 and

MVM), and 70% (between MVM and AAV). The homology between AAV and the other two viruses stops at this point (perhaps because the bulk of the AAV capsid sequence is encoded in a different ORF) but the homology between MVM and B19 continues for a significant distance, with 50% homology over 44 amino acids to nucleotide 2947 in B19 and 38% homology over 81 amino acids to nucleotide 3058 in B19. Although the function of this conserved region is currently obscure, its presence lends support to the suggestion that the sequence upstream of nucleotide 3125 in the major right-hand ORF of B19 encodes the VP-1 specific region, despite the fact that this amino acid sequence in B19 lacks the high concentration of basic residues and marked hydrophilic character seen in the VP-1 specific regions of the other autonomous parvoviruses analyzed to date.[56]

C. The Nonstructural Polypeptide Genes

The first parvoviral nonstructural protein was identified in CRFK cells infected with the ADV-G isolate of Aleutian disease virus.[47] This protein, which had an apparent molecular weight of 71 kdaltons and could be labeled in vitro with ^{32}P-orthophosphate, was compared to the two ADV capsid proteins (85 and 75 kdaltons) by two-dimensional chymotryptic map analysis and shown to have a very different amino acid sequence. Serum from mink with progressive Aleutian disease, but not normal, uninfected mink, had high levels of antibody against this protein, which were maintained throughout the course of the disease.

Subsequently, high molecular weight nonstructural proteins were also identified in the in vitro translation products of MVM, H-1, LuIII, and CPV-infected cell mRNA.[38,55,66] These polypeptides were shown to be of viral origin and to be synthesized from transcripts which encode the major open reading frame in the left half of the viral genome.[38,55] In MVM, this protein, designated NS-1, was shown to originate from the R1 transcript, which is synthesized from the promoter at map unit 4. The first AUG encountered downstream of the 5′ end of this transcript lies in the major open reading frame at position 261 in the MVM(p) sequence and prokaryotic expression studies confirm this AUG as the probable initiation codon for NS-1.[37,66] The whole of NS-1 appears to be encoded from a continuous sequence in this major ORF, terminating at nucleotide 2277 in MVM(p), since both alternative forms of the 46 to 48 map unit small splice occur downstream of this termination codon. Moreover, DNA fragments from sequences located upstream, but not from sequences located downstream, of this site were able to arrest the in vitro translation of the equivalent polypeptide, called NCVP1, encoded by H-1.[55] As seen in Figure 5B, antibodies raised against bacterial fusion proteins expressing the MVM(p) sequences in frame 3 between nucleotides 225 to 534 (fragment A), 591 to 957 (fragment B), and 1110 to 1628 (fragment C) all precipitate NS-1, confirming that this protein is the product of an R1 transcript encoding sequences throughout this entire region. An MVM NS-1 molecule specified in this way would contain 672 amino acids, which is somewhat less than would be expected from the apparent molecular weight of 83 kdaltons.

The NS-1 polypeptides of all the RV-like viruses show extensive antigenic conservation, in that sera from animals infected with any one of the serotypes will usually react with the NS-1 molecules, but rarely the capsid proteins, of any other member of the group. However, this antigenic conservation is not evenly spread throughout the molecule, but is particularly concentrated in a central "domain". Comparative sequence analysis has identified a region in the middle of the molecule which appears to be highly conserved between all members of the Parvoviridae examined to date. In a 405 base sequence from this region, corresponding to nucleotides 1428 through 1833 in MVM(p), the levels of nucleotide homology are 53% between MVM and AAV-2, 52% between MVM and B19, and 52% between B19 and AAV-2, while at the protein level these homologies are 51, 41, and 51%, respectively. Between members of the RV-like viruses these homologies are much more extensive, with MVM and H-1 being distinguished by a single amino acid, and FPV/CPV differing from

MVM and H-1 at six residues only (95.5% homology). Presumably, this sequence has been conserved because it forms a vital part of the active site of the molecule in one or more of the functions shared by this protein in all the parvoviruses.

Antibodies directed against bacterial proteins encoded by an MVM DNA sequence that includes part of this conserved region (nucleotides 1110 to 1638), cross-react well with the NS-1 molecules of H-1, LuIII, RV, PPV, and CPV synthesized in vitro or *in vivo*, but fail to react with the translation products of BPV or AAV-2 mRNA, showing that the most conserved residues are not immunodominant, at least in the denatured form of the protein.[66] Possibly one or more oligopeptides could be synthesized from this region which would be capable of eliciting antibodies that could identify the NS-1 proteins of the entire Parvoviridae and which, more importantly, might block whatever function is being performed by this protein domain. DNA sequence comparisons of the NS-1 protein genes of FPV/CPV and MVM downstream of this homology region (towards the carboxy-terminal of the polypeptide) reveal a substantial decline in the level of conservation (to around 57% if a few small gaps in the nucleotide line up are allowed), but the homology between these viruses still remains considerably higher than that seen between MVM, B19, and AAV-2, which is only about 16%. Moreover, antibodies directed against the amino-terminal domain of MVM NS-1 (between nucleotides 225 and 534) react efficiently with the homologous molecules from H-1 and LuIII, but entirely fail to react with the NS-1 of CPV.[66] Unfortunately, DNA sequence for this region of the CPV or FPV genome is not currently available, so it is not possible to say whether this lack of cross reactivity represents intense antigenic drift or the complete absence of an analogous region in the CPV molecule.

Antibodies directed against bacterial proteins expressing a sequence from the middle of the large ORF in the left end of the B19 genome (between nucleotides 1072 and 2044) recognized noncapsid proteins of 71, 63, and 52 kdaltons in the tissues of a human fetus infected transplacentally with B19,[46] but it is not yet clear whether these are primary translation products or processed forms of the nonstructural proteins. These antibodies did not recognize the NS polypeptides of the RV-like viruses or of BPV, even though the expressed peptide contained the highly conserved 135 amino acid sequence discussed previously, which lies between nucleotides 1390 and 1794 in the B19 sequence. This major ORF is preceded by a cluster of TATA boxes at the left end of the genome (at nucleotides 257, 319, 321, 323, and 412), one or more of which presumably function as part of a transcriptional promoter, but no CAAT sequences are present upstream of any of these putative promoters. An AUG at position 436 could initiate translation of the nonstructural proteins, which would be expected to give a molecule of 671 amino acids, terminating at nucleotide 2445, as a primary product. Such a protein might well migrate with an apparent molecular weight of 71 kdaltons, even though its calculated molecular weight would be 86 kdaltons. The left-hand ORF in B19 also contains an internal promoter-like sequence (with a TATA box at map unit 24, nucleotide 1225, and a CAAT box at 1195), which, if used to generate a transcript, could encode a protein initiating at an AUG codon at position 1288 and presumably coterminating with the larger polypeptide at 2445. Such a molecule would contain the carboxy-terminal 387 amino acids of NS-1 (including the 135 amino acid homology region) and would have a calculated molecular weight of 49 kdaltons. At present, however, we do not know if either the 63 or the 52 kdalton protein seen in vivo are produced in this way. Interestingly AAV-2 is known to have a functional promoter at around this position (map unit 18) in the NS-1 gene.

Preliminary sequence data shows that BPV has TATA boxes at map units 4 and 13, and a large open reading frame which extends through most of the left half of the genome and includes the NS-1 homology region detailed above.[57] Nevertheless, a high molecular weight nonstructural protein corresponding to this sequence has yet to be identified in vitro or in vivo. Instead, a major virally coded 27 kdalton nuclear phosphoprotein (designated NP1), which is not a constituent of the viral capsid, has been identified both in vivo and in the in vitro translation products of BPV-specific mRNA.[43,77] At present we do not know where this molecule is encoded in the viral DNA or if it is homologous to any of the proteins specified by the RV-like viruses. Interestingly, NP1 has recently been shown to bind DNA sequences from the 3′ end of the BPV genome.[78] Polypeptides of 25 and 22 kdaltons, which appear to be virally coded but which are not found in intact virions, were also identified in LPV-infected cells,[44] but at present it is not known if either of these correspond to the NP1 protein of BPV.

A second nonstructural protein (NS-2) has been identified in the in vitro translation products of H-1, MVM, and LuIII mRNA and in KRV-infected cells in culture.[66,37] As seen in Figure 5B, when synthesized in vitro from viral mRNA, this protein has an apparent molecular weight of 25 kdaltons. Prokaryotic expression studies reveal that the NS-2 molecule specified by MVM shares a common amino-terminal region with NS-1, encoded in frame 3, but utilizes a small block of alternative ORF in frame 2, located between nucleotides 2075 and 2291, to encode the carboxy-terminal half of the molecule (as depicted in Figure 5).[66] This suggests that NS-2 is the product of a 3.3 kbase R2 transcript which arises from the promoter at the left end of the genome and has two intervening sequences (map units 10 to 40 and 46 to 48) spliced out.

Recently Jongeneel et al.[72] isolated and sequenced a cDNA clone derived from MVM(i) mRNA in which nucleotide 515 was followed by nucleotide 1991, with the intervening sequence of 1475 bases deleted. Such a splice would join 84 amino-terminal residues encoded in frame 3 downstream of the AUG at 262 to 96 residues encoded in frame 2 between nucleotides 1992 and 2279, an arrangement which is clearly compatible with the organization of NS-2 as shown in Figure 5. This particular splice, which is summarized in Figure 6B, is somewhat unexpected since the splicing apparatus of the cell would have to ignore the potential 5′ splice sites at nucleotides 496, 520, and 541 shown in Figure 6B, which fit well to the consensus sequence determined from other eukaryotic proteins,[79] while recognizing the sequence AAA/GCAAGT, which is extremely unusual although not unique,[80] in having a C instead of a T at position 2 in the intron.

The cDNA clone isolated by Jongeneel also contained the minor splice which juxtaposes nucleotides 2280 and 2377 (MVM(p) nucleotide numbers), described as alternative splice (a) in Figure 6A. This would leave the NS-2 protein coding sequence in frame 2 beyond the minor splice for another six amino acids before terminating at an amber codon at nucleotide 2396. In the absence of this minor splice or if the alternative splice (b) as described in Figure 6A is used, the protein would still terminate seven residues downstream of nucleotide 2280 at position 2300. This is of interest because, as discussed previously, it has been shown that, while most R3 transcripts use this 5′ splice site, the 15% or so which encode VP-1, do not. This implies that NS-2 might exist in at least two forms with different carboxy-terminal hexapeptides. If the NS-2 transcripts use the splice sites described by Jongeneel et al.,[72] a protein starting at position 261 would contain 188 amino acids and have a calculated molecular weight of 25 kdaltons. NS-2 molecules translated from mRNAs which contain the map unit 46 to 48 splice alternative (a) would terminate in frame 2 at a single amber termination codon. However, after this single terminator, frame 2 is then open for another 96 codons to nucleotide 2667. As mentioned previously in the discussion of Figure 3, evidence from comparative sequence studies of MVM(p) and MVM(i) suggest that this alternative ORF downstream of the minor splice is used by the virus to encode essential protein, and it could well be that the translation apparatus of the cell is able to read through a proportion of these amber codons in a manner similar to that found, for instance, for the translation of the protease encoded by the gag-pol gene of murine leukemia virus.[81] Such natural suppression would give an NS-2 derivative molecule of 285 amino acids and a calculated molecular weight of 37 kdaltons. Although it has yet to be identified

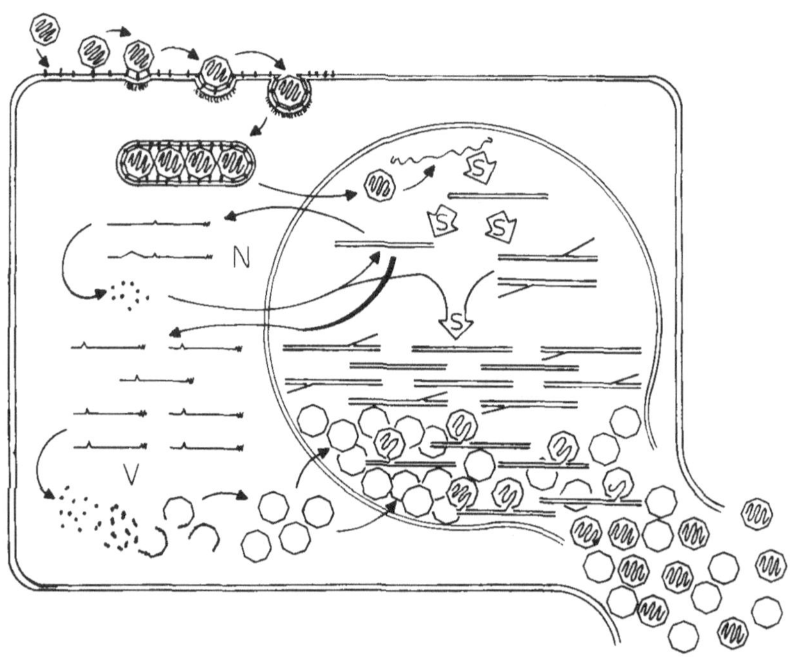

FIGURE 7. The parvovirus lytic growth cycle. Major features of the productive life cycle of an autonomous parvovirus, such as MVM(p), are shown in diagrammatic form. S denotes steps believed to be dependent upon host cell traverse of the S-phase. N denotes expression of the nonstructural proteins and up-regulation of the transcripts from P38, while V indicates the expression of R3 transcripts, followed by the synthesis of structural proteins, and assembly of empty capsids.

in vivo, the peptide originating from this open frame must confer unique properties on the whole protein, since it is exceedingly rich in leucine (20.8 mol %) and threonine (18.1 mol %), but contains only 11.1 mol % and 5.6 mol % of total basic or acidic amino acids, respectively. To date the existence of either the 6 kdalton protein or the 37 kdalton derivative of NS-2 as natural products of MVM gene expression has not been established.

Antibodies raised against the amino-terminal (nucleotides 225 to 534) or carboxy-terminal (nucleotides 2075 to 2291) domains of the MVM NS-2 fail to recognize homologous molecules in the translation products of CPV-infected cell mRNA,[66] and there is, as yet, no evidence that comparable proteins are encoded by FPV or CPV. Analysis of the DNA sequence of CPV or FPV (Figure 2) shows that, as in MVM, there is a small alternative ORF located just upstream of the map unit 46 to 48 splice (between nucleotides 288 and 566 in CPV), but this sequence terminates upstream of the 5′ splice site and, when compared to the amino acid sequence expressed in the carboxy-terminal half of NS-2, shows only very weak homology (a maximum of 28% over a region of 72 amino acids). Such observations indicate that the transcription patterns and organization of the nonstructural genes of FPV and CPV may be substantially different to those determined for MVM and H-1.

V. VIRUS-HOST CELL INTERACTIONS

A. The Productive Life Cycle

In common with most other nonenveloped viruses, the productive life cycle of the parvovirus, diagrammed in Figure 7, is always lytic. That is to say, the only known way in which progeny virus can be released from one infected cell and infect a neighboring cell, is by lysis and concomitant cell death. Productive infection is initiated by absorption of the virion to specific cell-surface receptors. The exact nature of these receptors is unknown, but *N*-acetyl neuraminic acid residues appear to play an essential role in the binding of MVM, since pretreating cells with neuraminidase prevents binding.[82] Pretreatment with trypsin also abolishes binding, suggesting that the receptors involve, or are displayed on, a protein backbone, but at present it is not established whether the receptor for a particular virus serotype is a single molecular species or a number of different molecules which all carry specific carbohydrate side chains. MVM is known to bind efficiently to a number of different cell types, from various species, and binding studies indicate that such cells frequently have between 10^5 and 5×10^5 specific binding sites per cell.[83,84] This remarkable abundance of receptor sites would tend to suggest that multiple protein species may be utilized, but expression of the receptor is nonetheless under developmental control, since some differentiated cell types, such as those of the B lymphocyte lineage, appear to lack receptors for MVM,[84] and are completely resistant to virus infection.

Very little is known about the next few stages of infection. As discussed previously, a proportion of the VP-2 molecules in full virions of the RV-like viruses are cleaved to yield VP-3 molecules, thus exposing a remarkably hydrophobic amino-terminal protein sequence near the virion exterior, which might conceivably be involved in the translocation of the virion across the plasma membrane. Empty particles cannot be cleaved at this site in VP-2, suggesting that they exhibit a different surface topography, but both full and empty particles compete equally for the receptor sites which lead to infection.[84] Internalization takes place through what appears to be coated pits,[83] and electron micrographs show cytoplasmic virus clustered within multivesicular bodies and heterophagosomes, with the viral particles either arrayed as a lining around the membrane or in paracrystalline aggregates.[85] The genome is then transported to the nucleus, prior to the first events of viral DNA replication. Whether it is uncoated in the cytoplasm, or arrives in the nucleus as an intact virion is not clear at present, but the latter possibility seems favored in view of the possible involvement of the incoming capsid in the initiation of viral gene expression to be discussed later.

During the course of a normal lytic infection, the MVM genome does not integrate into the host DNA,[86] although whether or not such events even occur during prolonged nonlytic interactions, such as the restrictive infections[87] to be described later, still needs further analysis. However, in the absence of a helper virus, AAV genomes integrate with high efficiency into host cell DNA,[88] and it will be interesting to determine whether autonomous viruses, such as B19, which have closely related terminal repeat sequences at each end of their genomes, rather than the unique termini found in MVM and H-1, share this integrative ability.

The processes of viral entry and accumulation in the nucleus can proceed synchronously in all the cells of an infected population irrespective of their position in the cell cycle.[5,6] In contrast, viral DNA replication and gene expression are entirely dependent upon one or more cellular functions expressed transiently during the S-phase of the cell cycle.[3-6] Since parvoviruses are unable to induce resting cells to enter S-phase but must wait for their host cell to embark on DNA replication of its own volition,[4] infecting randomly dividing cell populations leads to an asynchronous round of replication during which it is impossible to trace the ordered progression of specific viral activities. To alleviate this problem, a number of different regimens have been employed to induce a more or less synchronous cell cycle in parvovirus-infected cells,[5,6,85,89-91] but even these techniques, used individually, fail to give a population of cells which are sufficiently tightly coordinated as they enter S-phase to permit fine dissection of the viral life history. Better resolution can be achieved by employing a double block schedule, for example, by combining sequential isoleucine deprivation and inhibition of alpha DNA-polymerase activity.[37] Coupled with this type of synchronization, pulse-labeling and immunoprecipitation studies have shown that neither NS-1 nor the capsid polypeptides can be synthesized prior to S-phase. However, within 2 hr of entering S, *de novo* synthesis of NS-1, VP-1, and VP-2 can be detected, and this reaches maximal levels for all three proteins over the course of the next 6 hr, well before significant amplification of duplex viral DNA is observed.[37] Clearly an event which occurs at the very beginning of S-phase controls viral gene expression, and it is tempting to speculate that this event is the synthesis of the complementary DNA strand which thus provides a duplex template for viral transcription. There is considerable controversy, however, as to whether or not DNA-polymerase alpha is responsible for synthesis of this first complementary strand in vivo.[92,93]

If MVM(p) is used to infect highly synchronized cell populations, proteins encoded by both halves of the viral genome appear to be synthesized almost simultaneously within the first 2 hr of S. Since at least the majority of the capsid proteins are encoded by R3 transcripts arising from the promoter at map unit 38, while NS-1 is synthesized from an R1 transcript arising at map unit 4, this almost simultaneous early expression of all the major viral proteins appears to suggest that both promoters can be operational during this period. However, proof of this will require further analysis of the viral messenger RNAs, since it is formally possible that a small proportion of the capsid proteins synthesized early in infection could be derived from mRNAs arising at the promoter at map unit 4. Paradiso also found that *de novo* expression of all the major proteins of H-1 appeared to occur simultaneously,[94] but Molitor et al.[95] clearly demonstrated synthesis of the NS-1 protein of porcine parvovirus (PPV) 2 to 4 hr before they were able to detect capsid proteins. As in the previous studies, transcriptional data is not available, and detection of the protein species is indirect, relying on the ability of various antisera to precipitate the relevant proteins with equal efficiency. Nevertheless, it would seem that some viruses in some cell types can express and accumulate substantial levels of NS-1 prior to synthesis of detectable amounts of the capsid polypeptides. For MVM, H-1, and PPV, expression of the structural proteins appears to remain at peak levels long after synthesis of NS-1 has started to decline.

Parvovirus genomes appear to lack classical "early" and "late" transcription units, since it is doubtful if any RNA is made prior to the onset of viral complementary strand DNA synthesis. Moreover, for MVM and H-1, *de novo* synthesis of NS-1 and the capsid proteins is so tightly linked in time that, if these proteins are indeed the products of different transcription units early in infection, it seems unlikely that the activity of one promoter could be entirely dependent upon interaction with a protein specified by the other. Nevertheless, in DNA transfection studies synthesis of proteins encoded from the P38 promoter of H-1 was shown to be up-regulated 10- to 20-fold in NB cells by the simultaneous expression of NS-1.[67] It has also been suggested that NS-1 can down-regulate transcription from its own promoter,[68] but if so, kinetic studies of protein synthesis rates suggest that this is likely to be a very slow type of regulation which perhaps only operates when large concentrations of NS-1 are present in the cell.

When naked DNA fragments containing either of the MVM promoters were transcribed in nuclear extracts in vitro both promoters appeared to function with equal efficiency, whereas when nucleoprotein complexes isolated from infected A9 cells were similarly transcribed, the promoter at map unit 4 was ten times more active than that at map unit 38.[96] Although in vitro experiments are difficult to interpret, these studies reinforce the idea that in vivo the immediate protein environment of the transcription template is likely to influence profoundly the relative activity of the two promoters. An additional mechanism for modulating differential expression from the two promoters has been proposed by Ben-Asher and Aloni,[70] who identified a 142 nucleotide transcript synthesized from the P4 promoter and terminating immediately downstream of a short nucleotide sequence which is capable of forming alternative secondary structures. They proposed that the abundance of full length mRNAs arising from the P4 transcription unit might be regulated by interaction of a viral gene product with this sequence during RNA synthesis, favoring one of the two potential structures, and leading to either readthrough or premature termination, in a manner analogous to attenuation in prokaryotes.[97]

In vivo the NS-1 protein of MVM accumulates in the nucleus of the infected cell where, early in infection, it is found in two major forms, one of which co-migrates with its in vitro translation product (83 kdaltons) while the second, a phosphoprotein, migrates with a slightly higher apparent molecular weight (84 to 85 kdaltons).[98] Similar phosphorylated forms of NS-1 with abnormally high apparent molecular weights have also been identified in H-1 and PPV-infected cells and in each case phosphoserine was shown to be the predominant phosphorylated amino acid, although phosphothreonine residues were also detected in the NS-1 of PPV.[94,95] In PPV the unphosphorylated (or poorly phosphorylated) forms of NS-1 synthesized early in infection could be chased into the higher molecular weight form over the course of the next 1 to 2 hr.[95] The functional significance of this modification is unknown, but it appears to be associated with events occurring early in infection, since phosphorylated forms of NS-1 are often seen as soon as this molecule can be detected in the infected cell,[37,94] but are much less abundant late in infection.[37] A truncated form of NS-1 (around 60 kdaltons) also accumulates in the nucleus of most infected cells late in infection. It is not known if this abbreviated molecule plays any essential role in the biology of the virus or if it is an inevitable byproduct of one of the activities of full length NS-1, but for MVM(p) the same processed form appears to occur in every cell types examined so far.[37]

NS-1 probably carries out a number of quite different functions in vivo. Frameshift mutations introduced into the NS-1 coding sequence of an infectious plasmid clone of MVM DNA prevented excision and replication of the viral sequences following transfection of this construct into eukaryotic cells, but similar mutations in the capsid genes, while inhibiting single-strand DNA and virion synthesis, permitted high levels of duplex DNA replication.[99] This suggests that the NS-1 proteins play an essential role in the later stages of viral DNA replication, perhaps, in part, by cutting or nicking the DNA at specific sites, as discussed below. As previously mentioned these proteins also appear able to up-regulate transcription from a viral promoter at map unit 38 and, perhaps, to down-regulate the activity of both their own and a number of heterotogous eukaryotic promoters.[68] In highly synchronized, MVM-infected populations of A9 cells the rate of cellular DNA synthesis begins to decline 3 hr after the start of S-phase, although the peak of viral DNA replication is not destined to occur until much later. The decline in the rate of cellular DNA synthesis parallels the appearance of NS-1, although the capsid proteins are also fairly abundant in the cell by the time cellular DNA synthesis completely ceases.[37] The capsid polypeptides themselves do not, however, impair cell viability, since Pintel and colleagues were able to isolate stable cell lines which both expressed these proteins and assembled them into capsids.[71,100] However, repeated attempts to obtain cell lines expressing the nonstructural polypeptides have failed,[100,101] and Rhode has shown that the cotransfection of DNA encoding NS-1 together with constructs bearing various selectable genes into eukaryotic cells effectively abolishes the outgrowth of colonies expressing the selectable marker.[68] Thus NS-1 appears to prevent the long-term viability of its host cell, but whether it does this by a direct effect on DNA replication, or by disturbing cellular transcription remains to be determined. The amino-terminal domains of NS-1 and NS-2 share 84 amino acids of protein sequence, but they each have unique carboxy-terminal peptides.[66] The cellular location and function(s) of NS-2 have yet to be determined, but it seems possible that these molecules help to regulate one or more of the activities of NS-1 by competing for structures which interact with the amino-terminal sequence, while mediating their individual, and perhaps antagonistic, regulatory functions via their carboxy-terminal domains.

Viral DNA replication passes through three distinct phases: synthesis of a parental complementary DNA strand, amplification of duplex DNA forms, and excision and concomitant packaging of progeny, single-stranded viral DNA. Ward and Dadachanji[90] showed that in vivo full length complementary strands of MVM are synthesized from the 3′-hydroxyl of the input viral strands to yield covalently closed duplex molecules. Early in infection (6- or 8-hr postinfection in parasynchronized cells) much of the input-labeled, duplex DNA in the cell existed in this closed form, but by 16 hr postinfection nicks had been introduced into all but 18% of the converted input strands. Although synthesis of the complementary strand can occur as soon as the host cell enters S, major amplification of duplex DNA is delayed until late in S, and generally continues for several hours after uninfected control cells have ceased DNA replication.[37,89,91] The reasons for this delay are not understood, but it is tempting to speculate that this step in viral DNA replication requires an additional host cell function which is only available later in S. ^{32}P-labeled viral replicative form DNA extracted form asynchronous cultures of A9 cells 38 hr after infection with MVM(p) migrated in neutral agarose gels as single-stranded forms and as monomeric, dimeric, and tetrameric duplex molecules.[90] However, in denaturating gels these forms collapsed to single strands (64% of the labeled DNA) and monomer duplexes (32%), showing that the multimeric replicative form (RF) intermediates contained nicks which divided the DNA into unit length or twice unit length fragments. The restriction endonuclease EcoRI cuts the MVM genome twice, giving two easily distinguished terminal fragments, but EcoRI digestion of gel-purified monomer RF yielded two forms of each of the terminal fragments which had different electrophoretic mobilities. When the digestion mixture was heat-denatured, quench-cooled, and reelectrophoresed only the terminal fragments with the higher mobility from each end of the genome spontaneously reannealed, allowing them to migrate in their previous position, while the other form collapsed to the mobility expected for single-stranded molecules of this size. These were termed the "extended" and "turnaround" forms of the termini. The presence of spontaneously renaturing fragments from both the right and left ends of the genome is compatible with the idea that viral DNA synthesis occurs predominantly via a self priming mechanism. The larger, noncovalently closed forms of each terminus might then be explained by the introduction of a site-specific nick somewhere in the 3′ palindrome, and strand-displacement synthesis from the newly-exposed 3′-hydroxyl to complete the shorter strand, thus yielding two, complementary copies of the terminal hairpin. At the left end of the genome (the 3′ end of the viral strand) extended and turnaround forms were found with approximately equal frequency, but at the right end the extended form predominated. EcoRI digestion of dimer and tetramer RF molecules yielded similar extended and turnaround forms from both ends of the genome, but additional duplex fragments containing head-to-head copies of the left-hand EcoRI fragment (designated "dimer bridge") and tail-to-tail copies of the right-hand fragment ("tetramer bridge") were also identified. Upon heating these palindromic duplex fragments melted to co-migrate with turnaround fragments from the appropriate end of monomer RF. Thus dimers and tetramers are palindromic arrangements of unit length duplex DNA molecules arranged in head-to-head (viral 3′ to viral 3′) or tail-to-tail (viral 5′ to viral 5′) orientations, respectively. Pulse-chase experiments further revealed that these intermediates were metabolically active and eventually gave rise to progeny single-stranded viral DNA.[90]

The experimental data described here is drawn exclusively from early studies on MVM, and it is perhaps surprising that although many virus serotypes have been shown to have, for example, monomer and dimer duplex forms with a mixture of covalently closed and extended-form termini, detailed analyses of the structure and progression of replicative intermediates are not available for any other virus in the genus. Since this information is essential in the development of models for viral DNA replication, all such models rely heavily on the known structure of MVM and cannot be adequately cross checked by reference to other viruses. This is unfortunate since sequence analysis suggests that all members of the genus share a common ancestral form,[56] and it would be surprising if they employed very different replication strategies.

Despite considerable advances in the understanding of the molecular biology of eukaryotic DNA replication during the last few years, there have been relatively few observations made since 1978 which radically influence our view of the parvovirus replication process, and models available at that time[102] still form the basis for more recent proposals. However, two significant additions must be made in the light of DNA sequence studies into the structure of the termini of viral and replicative form DNA from MVM.[63] As discussed earlier the termini of all autonomous parvoviruses analyzed to date are imperfect palindromes, containing a few unpaired nucleotides. The presence of these unmatched residues allowed the identification of two sequence orientations termed "flip" and "flop", in the right-hand termini of both viral and RF molecules for MVM while only a single sequence was present at the left end of either DNA molecule.[63] Molecularly cloned plasmid forms of the MVM genome containing either of the two possible 5′ configurations gave wild-type virus with equal proportions of both forms upon transfection into susceptible cells,[99] showing that replication of a single 5′ DNA sequence inevitably regenerates both sequences. This, therefore, implies that in vivo the mechanism used to replicate the 5′ (right-hand) end of viral DNA must be different from, and less conservative than, that used to copy the 3′ (left-hand) end. Since the right-hand end of RF DNA was also shown to be 18 nucleotides longer than that of viral DNA[63] the nicking-site, and possibly also the enzymes, involved in generating the 5′ end of viral and RF DNA could be different. No differences in length between the left-hand ends of viral and RF DNA could be detected.

The limited genetic capacity of the parvoviral genome, and the fact that these viruses depend upon host function expressed transiently during the S-phase of the cell cycle, implies that for many aspects of their DNA replication parvoviruses depend upon the synthetic machinery of the cell. Replicating the ends of linear DNA molecules so that a single, perfect copy of each telomere is obtained after every round of synthesis presents particular problems, and since these rather complex structures are known to be of overriding importance for parvoviral DNA replication, most models focus on explaining their regeneration. A "rolling-hairpin" system of self-priming at the genomic termini coupled with a hairpin transfer mechanism, which adds the palindromic sequence at the 5′ end of one strand on to the 3′ end of its complementary strand,[102] form the basis for most models. A series of site-specific nicks and ligations must also be postulated and, since few details are available about either these or the enzymes and auxilliary proteins involved, an almost endless variety of marginally different mechanisms can be proposed.

There have been a number of reports of proteins associated both with the 5′ ends of RF DNA from a number of autonomous parvoviruses,[35,103-106] and with the 5′ ends of LuIII and MVM single-strand viral DNA.[34,35] Such proteins presumably mediate essential aspects of viral DNA replication. Iodinated DNA-protein complexes from MVM released a 60 to 65 kdalton protein upon DNAse digestion,[105] and Revie et al.[104] had previously estimated that a protein associated with H-1 RF molecules had a molecular weight of 60 to 80 kdaltons from changes in the buoyant density of RF DNA associated with the presence or absence of bound protein. However, more recent experiments by Paradiso[107] identified a 95 kdalton protein tightly associated with H-1 RF DNA and Wobbe et al.[106] demonstrated that both a 90 and a 40 kdalton species were specifically complexed with each of the terminal restriction fragments of RV RF DNA, but were not associated with internal restriction fragments from the same molecules. Importantly, it is clear that the terminal protein at the 5′ end of the viral strand in RF molecules, that is at the right-hand end, is attached at the extreme end of the extra 18 nucleotides which are present at the end of RF, but not viral DNA molecules. This implies that the terminal protein is either attached after the 18 nucleotides have been added, or is part of the mechanism behind their addition, as postulated below. Proteins reputed to be associated with viral single-strand DNA have proved to be even more elusive. Buoyant density studies[104] failed to reveal a protein attached to H-1 viral DNA, and those authors who have evidence for such a protein associated with MVM single-strands[35] report that the fraction of DNA complexed with protein varies substantially from preparation to preparation. Muller and Siegl[34] had previously shown that, unlike most mature forms of viral DNA, newly synthesized viral strands of LuIII prepared in an in vitro replication system were unable to penetrate a neutral agarose gel, suggesting that they were quantitatively complexed with protein. Thus, it may be that removal of any associated protein frequently occurs soon after packaging, but can vary in efficiency from one infection to another or from one cell type to another. At present, we do not know if the same polypeptides are associated with both viral and RF DNA, or what the exact functions of the viral strand terminal proteins are since, although it seems likely that they are involved in cutting the DNA at specific sites during replication, they may equally well influence packaging of the nascent single-strands or maturation of the newly assembled virion. As yet there is insufficient evidence to assess whether these terminal proteins are encoded by the virus or the host cell. A particularly interesting and provocative "modified rolling-hairpin" model for MVM replication has recently been put forward by Astell and colleagues[63,105] on the basis of the DNA sequence studies mentioned above, which provides a plausible explanation for the observed structures, and makes novel predictions about the nature of the enzymes used to generate the 3′ termini of viral strands. In discussing this overall scheme, presented in Figure 8, we have incorporated a number of additional steps as well as those central to the proposal of Astell et al.[63] According to this scheme, the 3′ terminal palindrome of the incoming V strand is used by a host polymerase to prime synthesis of a complementary strand, giving rise to a monomer duplex RF which is covalently closed at the left-hand end (step 1). Astell et al.[63] suggest that the 5′ hairpin is then displaced and copied (as in steps 4 and 5), although it is equally possible that the 5′ and 3′ ends of the molecule are first ligated (step 2), giving rise to a transient "collapsed-closed-circular" intermediate. We postulate that this intermediate could be the substrate for the sequence-specific nicking reaction, shown in step 3, which reopens the molecule 18 nucleotides inboard of the original viral strand 5′ end. For simplicity, we propose that this reaction is catalysed by the RF terminal protein Ⓣ, which remains covalently attached to the new 5′ end. The 3′ OH at this nick is now used as a primer for the strand-displacement synthesis shown in steps 4 and 5, to produce an RF molecule with a covalently closed turn-around terminus at the left-hand end, and an extended form right-hand end, including the terminal protein attached outboard of the extra 18-base-pair segment. At step 6, a rabbit-eared structure is formed and displacement synthesis occurs to produce a dimer RF (steps 7 and 8) in which the 3′ ends of the viral strands overlap in the center of the 10 kbase linear duplex molecule, while the 5′ ends of the viral strands are located at the molecular termini. As will be seen, involvement of a dimer intermediate at this point is essential for the regeneration of viral strands with a single sequence orientation at their 3′ termini. To achieve this, Astell and colleagues proposed that a nick is introduced, as shown in step 9, specifically in the parental V strand sequence opposite the 3′ end of the newly synthesized V strand, by a topoisomerase-like terminal protein, designated Ⓣ. Since it is crucial for the model that this nick is confined to the parental V strand, the authors propose that strand-preference may be determined by an asymmetry which results from mismatched nucleotides in the center of the main stem of the palindrome. The 3′ OH of this nick is then used for the extension of the parental V strand in a 5′ to 3′ direction, displacing the original 3′ palindrome of the parental V strand, as shown in step 10, enabling it to fold into a hairpin configuration, and regenerating an exact copy of this sequence in its original position. The authors propose that the same site-specific nickase remains covalently attached to the 5′ end of the nicked DNA during displacement and progressive reformation of the hairpin structure until it encounters a second copy of its DNA recognition sequence located at the 3′ end of the viral progeny strand (marked with an Ⓧ in the dimer produced by step 10). It then introduces a second nick at this new site and reseals a complete

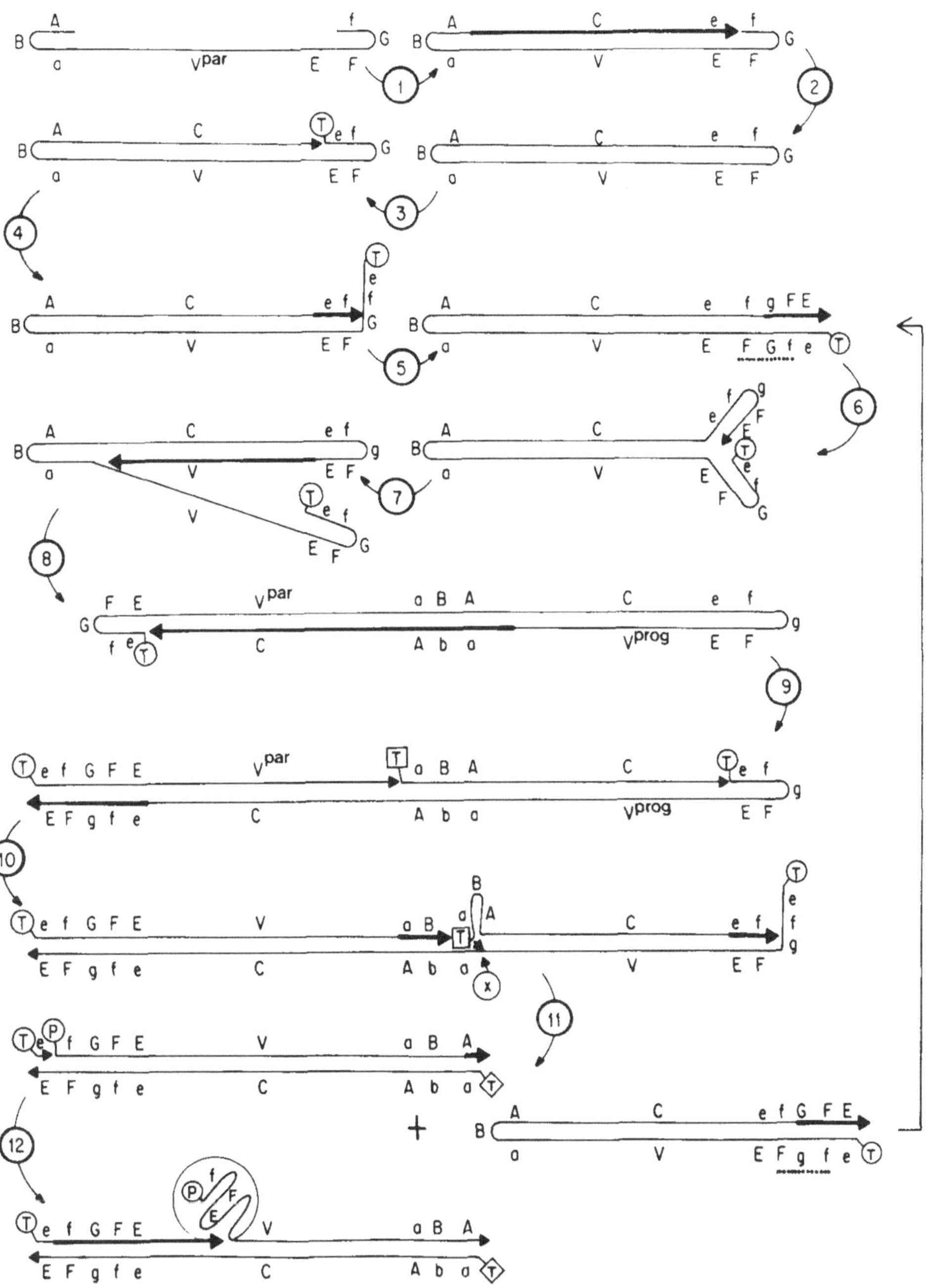

FIGURE 8. Modified "rolling hairpin" model for parvoviral DNA replication. The autonomous parvovirus genome is represented by a series of sequence blocks denoted by letters. Each upper and lower case letter denotes sequences complementary to one another. The 3′ end is shown on the left of the parental genome in the top left-hand corner, and is represented by A. The letters B and G denote the asymmetrical central portions of the left- and right-hand terminal palindromes, respectively. V^{par} and V^{prog} denote parental and progeny viral single-strand sequences, respectively, whereas the Complementary strand is represented by C. The 5′ end of mature virion DNA is represented by the letter f, while the letter e denotes the extra 18 base sequence found on the 5′ end of this strand in RF DNA molecules. The thick lines represent newly synthesized DNA, with the arrowhead denoting the 5′ to 3′ direction of DNA synthesis. The symbols Ⓣ and Ⓟ represent terminal protein/site-specific nucleases, and Ⓧ the topoisomerase-like reaction, discussed in the text.

3′ hairpin (step 11, right monomer), while itself being transferred to the 5′ end of the complementary strand (step 11, left monomer). At this point molecules such as that shown as the right-hand product of step 11 can recycle into the pool of replicating RF by undergoing steps 6 onward, as indicated by the long arrow at the right side of Figure 8. An important result of this synthetic loop is the observed generation of the two forms of the right-hand palindrome, designated FGf and Fgf at the 5′ end of progeny viral strands (underlined in Figure 8), and the conservation of the single sequence, designated ABa, at the 3′ ends of all mature viral strands. Molecules such as that shown as the left-hand product of step 11 can either regenerate monomers and dimers by rabbit-ear formation at either end of the genome, or go on to produce packaged genomes as shown in step 12. During the latter process we postulate that maturation of the correct viral 5′ ends occurs via site-specific nicking of the viral strand by a putative viral strand terminal protein Ⓟ at a position 18 nucleotides inboard of the RF terminal protein. While the 18 nucleotide fragment remaining attached to the RF terminal protein could then prime displacement synthesis of viral strands, the Ⓟ protein might direct packaging of displaced progeny viral strands into preformed capsids, a process which is known to be concomitant with and dependent upon viral strand DNA synthesis.[34,86] The Ⓟ protein would subsequently be cleaved off during final maturation of progeny virions.

As depicted in Figure 8, viruses such as MVM and H-1 package predominantly (> 99%) a single-sense DNA strand which is the complement of that encoded into RNA, and are thus, technically, negative-strand DNA viruses. Although this characteristic was initially presumed to be a diagnostic feature of the entire genus, more recent studies show that some autonomous parvoviruses, notably LuIII and B19, encapsidate both strands with approximately equal efficiency.[34,58,108,109] Since LuIII packages strands of both polarity, while H-1 grown in the same cell type only encapsidates the minus strand, the selection of strands appears to be predominantly virus-coded rather than host-specified, but the mechanism and structures responsible for this selection have yet to be determined. Relatively few of the preformed viral capsids are, however, destined to package DNA and mature into infectious virus particles, at least in in vitro virus-host cell systems. Ultimately a mixture of empty capsids and full virions is released from the cell, following nuclear degeneration and rupture of the plasma membrane by as yet undefined mechanisms.

B. Nonproductive Interactions

The requirement for host factors expressed only during S-phase and the inability of the virus to stimulate resting cells to initiate DNA synthesis have been established in cell culture systems as discussed previously, and are reflected in parvoviral disease of the whole animal as a preference for actively proliferating tissues as sites of attack. Not all virus-host cell interactions are productive, however, and some of these nonproductive interactions can still be lytic. A number of different types of nonproductive interactions have been investigated and in some cases one can learn much about particular virus-host relationships by studying systems in which the virus fails to grow.

As discussed previously, productive infection is initiated by adsorption of the virion to specific cell-surface receptors. The presence of functional receptors on the cell surface is apparently under developmental control, since some differentiated cell types lack them and are completely resistant to virus infection.[84] When the requirement for transition through S-phase is met in a receptor-positive, productive host cell, infection results in wholesale cell death. However, for many cell types, continued culture allows for the outgrowth of colonies of virus-resistant cells.[82,110] Resistant cells may have several different phenotypes depending upon which step of virus replication is affected. Common phenotypes include stable loss of expression of the cell-surface virus receptor,[110,111] inability to support viral transcription,[111] and inability to carry out an early step in either virus penetration or transport of input virus to the nucleus.[111] Although dissection of the host genes affected in such resistant cell variants is in its infancy, much information has already been gained about the cellular contribution to virus growth by studying the infection of naturally nonproductive cell types in vitro. Such virus-host interactions can usefully be divided into three categories; restrictive, abortive, and cryptic.

Restrictive infection, which is perhaps the best characterized of these, is exemplified by studies on the prototype (p) and immunosuppressive (i) variants of MVM. These virus strains differ from one another by being restricted for growth in one another's productive host cell type, fibroblasts, and T-lymphocytes, respectively.[87] The two viruses are serologically indistinguishable and have been designated allotropic variants of the MVM serotype, to indicate their distinct target cell specificities. Somatic cell hybrids between these two cell types are productive hosts for both viruses.[87] These results suggest that each restrictive host cell lacks a factor necessary for the productive replication of the restricted virus strain, and that these factors are expressed as a function of the differentiated state of the cell. Biochemical studies have demonstrated that these host factors operate at an intracellular level, and that both virus strains bind to, and compete with one another for, functional receptors on both cell types.[84,113] The first steps of virus growth are relatively unimpaired in restrictive infection,[84] and viral genomes are replicated to within 5 to 20% of the equivalent productive interaction levels[112,113] but no viral capsid proteins, progeny virions,[87] or single-stranded DNA[113] are detectable. Recent studies have shown the primary defect in restrictive infection to be a dramatic decrease in the initiation of viral transcription,[112] suggesting that the viral locus controlling strain-dependent target specificity, called the allotropic determinant, is a regulatory element which interacts with a developmentally regulated host factor to turn on viral gene expression. It is quite surprising, therefore, that genetic mapping of the allotropic determinant, by constructing recombinants between the two genomes, has located the fibrotropic determinant in MVM(p) between nucleotides 3522 and 3757, that is, in the VP-2 coding sequence within the capsid gene.[114] This 235 nucleotide long sequence, though only 4.5% of the genome, contains nearly 10% of the single-base differences between MVM(p) and MVM(i), and interestingly includes, in addition to 12 silent changes, one conservative amino acid difference and two of the three radical amino acid changes between the capsid genes of these two viruses.

In all restrictive virus-cell combinations examined in culture, there is a small subpopulation of cells, continually being replenished during the growth of the culture, which are able to support fully productive, lytic virus replication. The remainder of the culture restricts virus replication at the level of transcription and infection has little or no effect on cell growth rate or viability.[87] Infection of such cultures with a restricted virus can therefore result in a persistent carrier culture, which continually produces low levels of virus without apparent cytopathic effect. During passage of such cultures the low level of virus replication can give rise to a virus mutant, called an extended host range *(hr)* mutant, which has gained the ability to transcribe and replicate its genome in the formerly restrictive host cell type.[115] Such *hr* mutants may also preexist in a virus stock, and can be isolated directly as plaques on appropriate, normally restrictive cell monolayers. We have recently mapped the mutation in one such isolate, *hr* 101, a mutant of MVM(i) which is capable of growing productively in mouse fibroblasts, and found it to lie within the same small genomic region as the fibrotropic determinant of MVM(p), discussed above. Preliminary sequence data suggests that the mutation results in a radical amino acid change in the VP-1/VP-2 common region of the coat protein, although it is not the same change as either radical amino acid difference between MVM(p) and MVM(i).[116] Interestingly, Parrish and Carmichael[117] have reported the isolation and mapping of a CPV host-range mutant, selected by continued growth in feline cells, which has lost its ability to grow in canine cells. This mutation lies in the region of the CPV coat protein gene equivalent to that spanned by the two radical changes between MVM(p) and MVM(i), and also results in a radical amino acid substitution. It is not known, however, at what point in the normal replicative cycle the CPV mutant is blocked in canine cells. Taken together, these results strongly suggest that the capsid protein plays a role in the target cell specificity of the virus. Since in MVM this is mediated after entry of the virus but before expression of the coat protein gene, this implies that it is the incoming capsid which interacts with a developmentally regulated host cell factor to initiate viral transcription.

The appearance of *hr* mutants in a persistently infected restrictive cell culture results in a cytopathic crisis, with concomitant massive virus production. As this interaction is now productive, there is often a rapid selection of cellular variants which are resistant to the newly arisen virus mutant. It appears that many of these cellular variants can be restrictive for the new virus and the whole cycle can begin over again, and can apparently occur several times.[118]

In some types of nonproductive virus-cell interactions, especially those across species barriers (for instance, the infection of normal human fibroblast cell lines with rodent parvoviruses), significant expression of viral nonstructural and capsid antigens has been observed in the absence of detectable progeny virus production.[37,119,120] These interactions are termed abortive infections. In abortive infection the virus fails to establish its own DNA replication program, and the infection may or may not result in cell death, depending upon the particular combination of virus strain and cell line. All cells in the culture appear to be able to sustain viral transcription and antigen expression, often to levels found in productively infected cells.[37,119] This type of interaction is of particular interest since transformation of the cell by viral agents, such as SV40, convert the abortive interaction to a fully productive one. Dual SV40 and parvovirus infections are not productive, however, implying that transient SV40 early gene expression is not sufficient for reversal of the abortive phenotype, and that the establishment of the fully transformed state is required for these cells to support a productive parvovirus growth cycle.[121] Some mouse cell strains, both selected resistant mutants and apparently normal lines, possess this transformation-sensitive phenotype with respect to MVM infection.[23,122]

The final distinct type of interaction between virus and host cell occurs as a direct result of the requirement for the host to traverse S-phase and the inability of the virus to stimulate resting cells to do so. Thus, infection of naturally resting but otherwise productive host cells leads to a situation we denote cryptic infection. Examples of this are the infection of unstimulated porcine peripheral blood lymphocytes with PPV[123] or murine splenic T-cells with the immunosuppressive strain of MVM.[124] In these cases, activation of cryptically infected cells with concanavalin A or antigen results in active virus replication, leading to lysis of the activated cell.[123,124] Infection may then spread among other susceptible, cycling cells in the population.

The ability of a parvovirus to infect and kill its target cell, and produce infectious progeny virions, therefore depends upon the species and differentiated phenotype of the host cell, as well as its proliferative status. We can hope, then, that studies on B19 and other human parvoviruses, similar to those described in the preceding pages, will soon allow us to delineate the molecular mechanism behind parvoviral pathogenesis in man.

ACKNOWLEDGMENTS

We are very grateful to the numerous colleagues who communicated ideas and information prior to publication, and hope that these contributions are adequately referenced in the text. We also thank D. Greenberg for her help in processing the manuscript.

The authors were supported by Public Health Service Grants CA 16038 and CA 29303 from the National Cancer Institute.

REFERENCES

1. **Siegl, G., Bates, R. C., Berns, K. I., Carter, B. J., Kelly, D. C., Kurstak, E., and Tattersall, P.,** Characteristics and taxonomy of Parvoviridae, *Intervirology,* 23, 61, 1985.
2. **Berns, K. I., Ed.,** *The Parvoviruses,* Plenum Publ. New York, 1984.
3. **Tennant, R. W., Layman, K. R., and Hand, R. E., Jr.,** Effect of cell physiological state on infection by rat virus, *J. Virol.,* 4, 872, 1969.
4. **Tattersall, P.,** Replication of the parvovirus MVM. Dependence of virus multiplication and plaque formation on cell growth, *J. Virol.,* 10, 586, 1972.
5. **Siegl, G. and Gautschi, M.,** The multiplication of parvovirus LuIII in a synchronized culture system. I. Optimum conditions for virus replication, *Arch. Gesamte Virusforsch.,* 40, 105, 1973.
6. **Rhode, S. L., III,** Replication process of the parvovirus H-1. I. Kinetics in a parasynchronous cell system, *J. Virol.,* 11, 856, 1973.
7. **Hampton, E. G.,** H-1 growth in synchronized rat embryo cells, *Can. J. Microbiol.,* 16, 266, 1970.
8. **Margolis, G. and Kilham, L.,** Problems of human concern arising from animal models of intrauterine and neonatal infections due to viruses: a review. II. Pathologic studies, *Prog. Med. Virol.,* 20, 144, 1975.
9. **Kilham, L. and Margolis, G.,** Problems of human concern arising from animal models of intrauterine and neonatal infections due to viruses: a review. I. Introduction and virologic studies, *Prog. Med. Virol.,* 20, 113, 1975.
10. **Siegl, G.,** Biology and pathogenicity of autonomous parvoviruses, in *The Parvoviruses,* Berns, K. I., Ed., Plenum Press, New York, 1984, 297.
11. **Lipton, H. L. and Johnson, R. T.,** The pathogenesis of rat virus infections in the newborn hamster, *Lab. Invest.,* 27, 508, 1972.
12. **Siegl, G.,** Canine parvovirus: origin and significance of a "new" pathogen, in *The Parvoviruses,* Berns, K. I., Ed., Plenum Press, New York, 1984, 363.
13. **Porter, D. D. and Cho, H. J.,** Aleutian Disease of mink: a model for persistent infection, in *Comprehensive Virology,* Vol. 16, Fraenkel-Conrat, H. and Wagner, R. R., Eds., Plenum Press, New York, 1980, 233.
14. **Mohanty, S. B. and Bachmann, P. A.,** Susceptibility of fertilized mouse eggs to minute virus of mice, *Infect. Immun.,* 9, 762, 1974.
15. **Miller, R. A., Ward, D. C., and Ruddle, F. H.,** Embryonal carcinoma cells (and their somatic cell hybrids) are resistant to infection by the murine parvovirus MVM, which does infect other teratocarcinoma-derived cell lines, *J. Cell. Physiol.,* 91, 393, 1977.
16. **Tattersall, P.,** Susceptibility to minute virus of mice as a function of host cell differentiation, in *Replication of Mammalian Parvoviruses,* Ward, D. C. and Tattersall, P., Eds., Cold Spring Harbor Laboratory, Cold Spring Harbor, N.Y., 1978, 131.
17. **Kilham, L. and Olivier, L. J.,** A latent virus of rats isolated in tissue culture, *Virology,* 7, 428, 1959.
18. **Toolan, H. W.,** Studies on a viral agent associated with human tumors, *Proc. Am. Assoc. Cancer Res.,* 3, 273, 1961.
19. **Toolan, H. W.,** Lack of oncogenic effect of the H-viruses, *Nature (London),* 214, 1036, 1967.
20. **Toolan, H. W. and Ledinko, N.,** Inhibition by H-1 virus of the incidence of tumors produced by adenovirus 12 in hamsters, *Virology,* 35, 475, 1968.
21. **Bergs, V. V.,** Rat virus-mediated suppression of leukemia induction by Moloney virus in rats, *Cancer Res.,* 29, 1669, 1969.
22. **Toolan, H. W., Rhode, S. L., III, and Gierthy, J. F.,** Inhibition of 7,12-dimethylbenz(a)anthracene-induced tumors in syrian hamsters by prior infection with H-1 parvovirus, *Cancer Res.,* 42, 2552, 1982.
23. **Mousset, S. and Rommelaere, J.,** Minute virus of mice inhibits cell transformation by simian virus 40, *Nature (London),* 300, 537, 1982.
24. **Campbell, D. A., Staal, S. P., Manders, E. K., Bonnard, G. D., Oldham, R. K., Salzman, L. A., and Herberman, R. B.,** Inhibition of in vitro lymphoproliferative responses by in vivo passaged rat 13762 mammary adenocarcinoma cells. III. Evidence that Kilham rat virus is responsible for the inhibitory effect, *Cell. Immunol.,* 33, 378, 1977.
25. **Lum, G. S. and Schreiner, A. W.,** Study of a virus isolated from a chloroleukemic Wister rat, *Cancer Res.,* 23, 1742, 1963.
26. **Bonnard, G. D., Manders, E. K., Campbell, D. A., Herberman, R. B., and Collins, M. J.,** Immunosuppressive activity of a subline of the mouse EL-4 lymphoma. Evidence for minute virus of mice causing the inhibition, *J. Exp. Med.,* 143, 187, 1976.
27. **McMaster, G. K., Beard, P., Engers, H. D., and Hirt, B.,** Characterization of an immunosuppressive parvovirus related to the minute virus of mice, *J. Virol.,* 38, 317, 1981.
28. **Engers, H. D., Louis, J. A., Zubler, R. H., and Hirt, B.,** Inhibition of T cell-mediated functions by MVM(i), a parvovirus closely related to minute virus of mice, *J. Immunol.,* 127, 2280, 1981.

29. **Tattersall, P.**, Parvovirus protein structure and virion maturation, in *Replication of Mammalian Parvoviruses,* Ward, D. C. and Tattersall, P., Eds., Cold Spring Harbor Laboratory, Cold Spring Harbor, N.Y., 1978, 53.
30. **Ward, D. C. and Tattersall, P.**, Minute virus of mice, in *The Mouse in Biomedical Research,* Vol. 2, Foster, H. L., Small, J. D., and Fox, J. G., Eds., Academic Press, New York, 1982, 313.
31. **Paradiso, P. R.**, Infectious process of the parvovirus H-1: correlation of protein content, particle density, and viral infectivity, *J. Virol.*, 39, 800, 1981.
32. **Faust, E. A. and Ward, D. C.**, Incomplete genomes of the parvovirus minute virus of mice: selective conservation of genome termini, *J. Virol.*, 32, 276, 1979.
33. **Kelly, D. C. and Elliot, R. M.**, Polyamines contained by two Densonucleosis viruses, *J. Virol.*, 21, 408, 1977.
34. **Muller, D-E and Siegl, G.**, Maturation of parvovirus LuIII in a subcellular system. I. Optimal conditions for in vitro synthesis and encapsidation of viral DNA, *J. Gen. Virol.*, 64, 1043, 1983.
35. **Chow, M., Bodnar, J. W., Polvino-Bodnar, M., and Ward, D. C.**, Identification and characterization of a protein covalently bound to DNA of minute virus of mice, *J. Virol.*, 57, 1094, 1986.
36. **Chow, M. B.**, DNA Replication of Minute Virus of Mice, A Parvovirus, Ph.D. thesis, Yale University, New Haven, Conn., 1981.
37. **Tattersall, P. and Cotmore, S. F.**, unpublished observations.
38. **Cotmore, S. F., Sturzenbecker, L. J., and Tattersall, P.**, The autonomous parvovirus MVM encodes two nonstructural proteins in addition to its capsid polypeptides, *Virology,* 129, 333, 1983.
39. **Clinton, G. M. and Hayashi, M.**, The parvovirus MVM: a comparison of heavy and light particle infectivity and their density conversion in vitro, *Virology,* 74, 57, 1976.
40. **Tattersall, P., Shatkin, A. J., and Ward, D. C.**, Sequence homology between the structural polypeptides of minute virus of mice, *J. Mol. Biol.*, 111, 375, 1977.
41. **Peterson, J. L., Dale, R. M. K., Karess, R., Leonard, D., and Ward, D. C.**, Comparison of parvovirus structural proteins: evidence for post-translational modification, in *Replication of Mammalian Parvoviruses,* Ward, D. C. and Tattersall, P., Eds., Cold Spring Harbor Laboratory, Cold Spring Harbor, N.Y., 1978, 431.
42. **Paradiso, P. R., Williams, K. R., and Constantino, R. L.**, Mapping of the amino terminus of the H-1 parvovirus major capsid protein, *J. Virol.*, 52, 77, 1984.
43. **Lederman, M., Bates, R. C., and Stout, E. R.**, In vitro and in vivo studies of bovine parvovirus proteins, *J. Virol.*, 48, 10, 1983.
44. **Matsunaga, Y. and Matsuno, S.**, Structural and nonstructural proteins of a rabbit parvovirus, *J. Virol.*, 45, 627, 1983.
45. **Matsunaga, Y., Matsuno, S., and Mukoyama, J.**, Isolation and characterization of a parvovirus of rabbits, *Infect. Immun.*, 18, 495, 1977.
46. **Cotmore, S. F., McKie, V. C., Anderson, L. J., Astell, C. R., and Tattersall, P.**, Identification of the major structural and non-structural proteins encoded by the human parvovirus B19 and mapping of their genes by procaryotic expression of isolated genomic fragments, *J. Virol.*, 60, 548, 1986.
47. **Bloom, M. E., Race, R. E., and Wolfinbarger, J. B.**, Identification of a nonvirion protein of Aleutian Disease virus: mink with Aleutian Disease have antibody to both virion and nonvirion proteins, *J. Virol.*, 43, 608, 1982.
48. **Aasted, B., Race, R. E., and Bloom, M. E.**, Aleutian Disease virus, a parvovirus, is proteolytically degraded during in vivo infection in mink, *J. Virol.*, 51, 7, 1984.
49. **Parrish, C. R. and Carmichael, L. E.**, Antigenic structure and variation of canine parvovirus type-2, feline panleukopenia virus, and mink enteritis virus, *Virology,* 129, 401, 1983.
50. **Burtonboy, G., Delferiere, N., Surleraux, M., and Bodeus, M.**, Epitopes and Neutralization of CPV 2, EMBO Parvovirus Workshop, Grangeneuve, Switzerland, 1985.
51. **Bourguignon, G. J., Tattersall, P. J., and Ward, D. C.**, DNA of minute virus of mice: self-priming, non-permuted, single-stranded genome with a 5′ terminal hairpin duplex, *J. Virol.*, 20, 290, 1976.
52. **Astell, C. R., Thomson, M., Merchlinsky, M., and Ward, D. C.**, The complete DNA sequence of minute virus of mice, an autonomous parvovirus, *Nucleic Acids Res.*, 11, 999, 1983.
53. **Astell, C. R., Gardiner, E. M., and Tattersall, P. J.**, The DNA sequence of the lymphotropic variant of minute virus of mice, MVM(i), and its comparison to that of the fibrotropic prototype strain, *J. Virol.*, 57, 656, 1986.
54. **Sahli, R., McMaster, G. K., and Hirt, B.**, DNA sequence comparison between two tissue-specific variants of the autonomous parvovirus, minute virus of mice, *Nucleic Acids Res.*, 13, 3617, 1985.
55. **Rhode, S. L., III and Paradiso, P. R.**, Parvovirus genome: nucleotide sequence of H-1 and mapping of its genes by hybrid-arrested translation, *J. Virol.*, 45, 173, 1983.
56. **Shade, R. O., Blundell, M. C., Cotmore, S. F., Tattersall, P., and Astell, C. R.**, Nucleotide sequence and genome organization of human parvovirus B19 isolated from the serum of a child during aplastic crisis, *J. Virol.*, 58, 921, 1986.

57. **Schull, B. C., Lederman, M., Chen, K. C., Moses, E. A., Stout, E. R., and Bates, R. C.,** Sequence and Expression of Cloned Bovine Parvovirus DNA, EMBO Parvovirus Workshop, Grangeneuve, Switzerland, 1985.
58. **Cotmore, S. F. and Tattersall, P.,** Characterization and molecular cloning of a human parvovirus genome, *Science,* 226, 1161, 1984.
59. **Banerjee, P. T., Olson, W. H., Allison, D. P., Bates, R. C., Snyder, C. E., and Mitra, S.,** Electron microscopic comparison of the sequences of single-stranded genomes of mammalian parvoviruses by heteroduplex mapping, *J. Mol. Biol.,* 166, 257, 1983.
60. **Clewley, J. P.,** Cloning of DNA from a Putative Parvovirus from Human Faeces, EMBO Parvovirus Workshop, Grangeneuve, Switzerland, 1985.
61. **Astell, C. R., Smith, M., Chow, M. B., and Ward, D. C.,** Structure of the 3′hairpin termini of four rodent parvovirus genomes: nucleotide sequence homology at origins of DNA replication, *Cell,* 17, 691, 1979.
62. **Rhode, S. L., III and Klaassen, B.,** DNA sequence of the 5′ terminus containing the replication origin of parvovirus replicative form DNA, *J. Virol.,* 41, 990, 1982.
63. **Astell, C. R., Chow, M. B., and Ward, D. C.,** Sequence analysis of the termini of virion and replicative forms of minute virus of mice DNA suggests a modified rolling hairpin model for autonomous parvovirus DNA replication., *J. Virol.,* 54, 171, 1985.
64. **Merchlinsky, M. J., Tattersall, P. J., Leary, J. J., Cotmore, S. F., Gardiner, E. M., and Ward, D. C.,** Construction of an infectious molecular clone of the autonomous parvovirus minute virus of mice, *J. Virol.,* 47, 227, 1983.
65. **Boissy, R. and Astell, C. R.,** An *Escherichia coli* recBCsbcBrecF host permits the deletion-resistant propagation of plasmid clones containing the 5′-terminal palindrome of minute virus of mice, *Gene,* 35, 179, 1985.
66. **Cotmore, S. F. and Tattersall, P.,** Organization of the non-structural genes of the autonomous parvovirus minute virus of mice, *J. Virol.,* 58, 724, 1986.
67. **Rhode, S. L., III,** Trans-activation of parvovirus P38 promoter by the 76K noncapsid protein, *J. Virol.,* 55, 886, 1985.
68. **Rhode, S. L., III,** Regulation of Parvovirus Transcription by Non-capsid Protein, EMBO Parvovirus Workshop, Grangeneuve, Switzerland, 1985.
69. **Pintel, D., Dadachanji, D., Astell, C. R., and Ward, D. C.,** The genome of minute virus of mice, an autonomous parvovirus, encodes two overlapping transcription units, *Nucleic Acids Res.,* 11, 1019, 1983.
70. **Ben-Asher, E. and Aloni, Y.,** Transcription of minute virus of mice, an autonomous parvovirus, may be regulated by attenuation, *J. Virol.,* 52, 266, 1984.
71. **Labieniec-Pintel, L. and Pintel, D.,** The minute virus of mice (MVM) P39 transcription unit can encode both capsid proteins, *J. Virol.,* 57, 1163, 1986.
72. **Jongeneel, C. V., McMaster, G., Sahli, R., and Hirt, B.,** Analysis of the Genome Organization of MVM by cDNA Cloning, EMBO Parvovirus Workshop, Grangeneuve, Switzerland, 1985.
73. **Brown, D. D. and Salzman, L. A.,** Sequence homology between the structural proteins of Kilham rat virus, *J. Virol.,* 49, 1018, 1984.
74. **Jongeneel, C. V.,** personal communication.
75. **Morgan, W. R. and Ward, D. C.,** personal communication.
76. **Becerra, S. P., Rose, J. A., Hardy, M., Baroudy, B. M., and Anderson, C. W.,** Direct mapping of adeno-associated virus capsid proteins B and C: a possible ACG initiation codon, *Proc. Natl. Acad. Sci. U.S.A.,* 82, 7919, 1985.
77. **Lederman, M., Patton, J. T., Stout, E. R., and Bates, R. C.,** Virally coded noncapsid protein associated with bovine parvovirus infection, *J. Virol.,* 49, 315, 1984.
78. **Lederman, M., Bates, R. C., and Stout, E. R.,** DNA Binding Properties of Bovine Parvovirus Proteins, EMBO Parvovirus Workshop, Grangeneuve, Switzerland, 1985.
79. **Mount, S. M.,** A catalogue of splice junction sequences, *Nucleic Acids Res.,* 10, 459, 1982.
80. **Fischer, H. D., Dodgson, J. B., Hughes, S., and Engel, J. D.,** An unusual 5′ splice sequence is efficiently utilized in vivo, *Proc. Natl. Acad. Sci. U.S.A.,* 81, 2733, 1984.
81. **Yoshiyuki, Y., Katoh, I., Copeland, T. D., and Oroszlan, S.,** Murine leukemia virus protease is encoded by the gag-pol gene and is synthesized through suppression of an amber termination codon, *Proc. Natl. Acad. Sci. U.S.A.,* 82, 1618, 1985.
82. **Stanick, D., Moir, R., Cotmore, S. F., and Tattersall, P.,** unpublished observations.
83. **Linser, P., Bruning, H., and Armentrout, R. W.,** Specific binding sites for a parvovirus, minute virus of mice, on cultured mouse cells, *J. Virol.,* 24, 211, 1977.
84. **Spalholz, B. A. and Tattersall, P.,** Interaction of minute virus of mice with differentiated cells: strain-dependent target cell specificity is mediated by intracellular factors, *J. Virol.,* 46, 937, 1983.
85. **Richards, R., Linser, P., and Armentrout, R. W.,** Kinetics of assembly of a parvovirus, minute virus of mice, in syncronized rat brain cells, *J. Virol.,* 22, 778, 1977.

86. **Richards, R. G. and Armentrout, R. W.,** Early events in parvovirus replication: lack of integration by minute virus of mice into host cell DNA, *J. Virol.*, 30, 397, 1979.
87. **Tattersall, P. and Bratton, J.,** Reciprocal productive and restrictive virus-cell interactions of immunosuppressive and prototype strains of minute virus of mice, *J. Virol.*, 46, 944, 1983.
88. **Cheung, A., Hoggan, M. D., Hauswirth, W. W., and Berns, K. I.,** Integration of the adeno-associated virus genome into cellular DNA in latently infected human Detroit 6 cells, *J. Virol.*, 33, 739, 1980.
89. **Parris, D. S. and Bates, R. C.,** Effect of bovine parvovirus replication on DNA, RNA, and protein synthesis in S-phase cells, *Virology*, 73, 72, 1976.
90. **Ward, D. C. and Dadachanji, D.,** Replication of minute virus of mice DNA, in *Replication of Mammalian Parvoviruses*, Ward, D. C. and Tattersall, P., Eds., Cold Spring Harbor Laboratory, Cold Spring Harbor, N.Y., 1978, 297.
91. **Hardt, N., Dinsart, C., Spadari, S., Pedrali-Noy, G., and Rommelaere, J.,** Interrelation between viral and cellular DNA synthesis in mouse cells infected with the parvovirus minute virus of mice, *J. Gen. Virol.*, 64, 1991, 1983.
92. **Kollek, R., Tseng, B. Y., and Goulian, M.,** DNA polymerase requirements for parvovirus H-1 DNA replication in vitro, *J. Virol.*, 41, 982, 1982.
93. **Faust, E. A. and Rankin, C. D.,** In vitro conversion of MVM parvovirus single-stranded DNA to the replicative form by DNA polymerase a from Ehrlich ascites tumor cells, *Nucleic Acids Res.*, 10, 4181, 1982.
94. **Paradiso, P. R.,** Identification of multiple forms of the noncapsid parvovirus protein NCVP1 in H-1 parvovirus-infected cells, *J. Virol.*, 52, 82, 1984.
95. **Molitor, T. W., Joo, H. S., and Collett, M. S.,** Identification and characterization of a porcine parvovirus nonstructural polypeptide, *J. Virol.*, 55, 554, 1985.
96. **Beard, P., Bruggmon, H., and Doerig, C.,** Transcription of MVM nucleoprotein complexes, EMBO Parvovirus Workshop, Grangeneuve, Switzerland, 1985.
97. **Yanofsky, C.,** Attenenuation in the control expression of bacterial operans, *Nature (London)*, 289, 751, 1981.
98. **Cotmore, S. F. and Tattersall, P.,** The NS-1 polypeptide of the autonomous parvovirus MVM is a nuclear phosphoprotein, *Virus Res.*, 4, 243, 1986.
99. **Merchlinsky, M. J.,** Studies on MVM DNA Replication Using an Infectious Molecular Clone; Ph.D. thesis, Yale University, New Haven, Conn., 1984.
100. **Pintel, D., Merchlinsky, M. J., and Ward, D. C.,** Expression of minute virus of mice structural proteins in murine cell lines transformed by bovine papillomavirus-minute virus of mice plasmid chimera, *J. Virol.*, 52, 320, 1984.
101. **Moir, R., Cotmore, S. F., and Tattersall, P.,** unpublished observations.
102. **Tattersall, P. and Ward, D. C.,** Rolling hairpin model for replication of parvovirus and linear chromosomal DNA, *Nature (London)*, 263, 106, 1976.
103. **Hayward, G. S., Bujard, H., and Gunther, M.,** Three distinct replicative forms of Kilham-rat-virus DNA, in *Replication of Mammalian Parvoviruses*, Ward, D. C. and Tattersall, P., Eds., Cold Spring Harbor Laboratory, Cold Spring Harbor, N.Y., 1978, 327.
104. **Revie, D., Tseng, B. Y., Graftsom, R. H., and Goulian, M.,** Covalent association of protein with replicative form DNA of parvovirus Hl, *Proc. Natl. Acad. Sci. U.S.A.*, 76, 5539, 1979.
105. **Astell, C. R., Thomson, M., Chow, M. B., and Ward, D. C.,** Structure and replication of minute virus of mice DNA, *Cold Spring Harbor Symp. Quant. Biol.*, 47, 751, 1983.
106. **Wobbe, C. R. and Mitra, S.,** Proteins tightly associated with the termini of replicative form DNA of Kilham rat virus, an autonomous parvovirus, *Proc. Natl. Acad. Sci. U.S.A.*, 82, 8335, 1985.
107. **Paradiso, P.,** personal communication.
108. **Summers, J., Jones, S. E., and Anderson, M. J.,** Characterization of the genome of the agent of erythrocyte aplasia permits its classification as a human parvovirus, *J. Gen. Virol.*, 64, 2527, 1983.
109. **Bates, R. C., Snyder, C. E., Banerjee, P. T., and Mitra, S.,** Autonomous parvovirus LuIII encapsidates equal amounts of plus and minus DNA strands, *J. Virol.*, 49, 319, 1984.
110. **Linser, P. and Armentrout, R. W.,** Binding of minute virus of mice to cells in culture, in *Replication of Mammalian Parvoviruses*, Ward, D. C. and Tattersall, P., Eds., Cold Spring Harbor Laboratory, Cold Spring Harbor, N.Y., 1978, 151.
111. **Stanick, D. and Tattersall, P.,** unpublished observations.
112. **Spalholz, B. A., Ward, D. C., and Tattersall, P.,** manuscript in perparation.
113. **Spalholz, B. A., Bratton, J., Ward, D. C., and Tattersall, P.,** Intracellular restriction of parvovirus replication in differentiated cells, in *Tumor Viruses and Differentiation*, Levine, A. and Scolnick, E., Eds., Alan R. Liss, New York, 1983, 315.
114. **Gardiner, E. M. and Tattersall, P.,** unpublished observations.
115. **Ron, D., Tattersall, P., and Tal, J.,** Formation of a host range mutant of the lymphotropic strain of minute virus of mice during persistent infection in mouse L cells, *J. Virol.*, 52, 63, 1984.

116. **Moir, R., Gardiner, E. M., and Tattersall, P.,** unpublished observations.
117. **Parrish, C. R. and Carmichael, L. E.,** Characterization and recombination mapping of an antigenic and host range mutation of canine parvovirus, *Virology,* 148, 121, 1986.
118. **Ron, D. and Tal, J.,** Coevolution of cells and virus as a mechanism for the persistence of lymphotropic minute virus of mice in L-cells, *J. Virol.,* 55, 424, 1985.
119. **de Foresta, F., Chen, Y. Q., Hertoghs, J., Avalosse, B., Duponchel, N., Cornelis, J. J., and Rommelaere, J.,** Transformation by SV40 Sensitizes Human Skin and Lung Fibroblasts to the Cytopathic Effect of Parvovirus H-1, EMBO Parvovirus Workshop, Grangeneuve, Switzerland, 1985.
120. **de Foresta, F., Deleys-Hertoghs, J., Cornelis, J., Klein, B., and Rommelaere, J.,** Transformation by SV40 virus sensitizes fibroblasts of human skin to the lytic action of H-1 parvovirus, *Crit. Rev. Soc. Biol. Paris,* 179, 276, 1985.
121. **Tattersall, P. and Bratton, J.,** unpublished observations.
122. **Rommelaere, J.,** personal communication.
123. **Paul, P. S., Mengeling, W. L., and Brown, T. T., Jr.,** Replication of porcine parvovirus in peripheral blood lymphocytes, monocytes, and peritoneal macrophages, *Infect. Immun.,* 25, 1003, 1979.
124. **Molitor, T. W., Winograd, D., Smith, A. L., and Tattersall, P.,** unpublished observations.
125. **Tattersall, P., Cawte, P. J., Shatkin, A. J., and Ward, D. C.,** Three structural polypeptides coded for by minute virus of mice, a parvovirus, *J. Virol.,* 20, 273, 1976.

Chapter 3

PATTERNS OF PARVOVIRUS DISEASE IN ANIMALS

Günter Siegl

TABLE OF CONTENTS

I. INTRODUCTION

It has only been a relatively short period of time since the medical profession has taken interest in the nature, the characteristics, and the pathogenic potential of parvoviruses. For years, these viruses have been regarded solely as study objects suitable for the molecular virologist or, at best, for veterinary microbiologists and pathologists. This, however, has not always been the case. At the very beginning of parvovirus research, the frequent recovery of these agents from tumor bearing rodents,[1] from transplantable human tumors,[2] or from permanent human cell lines of tumor origin[3,4] appeared to point to an active role of parvoviruses in malignant transformation of animal and human cells. Studies in the first half of the 1960s showed that this is not true and soon revealed the molecular basis for the predilection of parvoviruses for rapidly proliferating cells.[5] The very same experiments also led to the discovery of the role of parvoviruses as agents of specific natural diseases.[6-9] Feline cerebellar ataxia and feline panleukopenia, both known as ravaging diseases in feline populations for decades (see Reference 10) were the first diseases shown to be caused by one distinct parvovirus. Later on, a broad spectrum or syndromes like enteritis, hepatitis, myocarditis, immune complex disease, and congenital malformation could be related to parvovirus infections in other animal species.

Enteritis, hepatitis, ataxia, myocarditis, and congenital malformations are also well known in man. Therefore, much time has been spent throughout the past two decades to establish links between parvovirus infections and such syndromes in the human patient. However, as has been characteristic for many of the animal parvoviruses, discovery of the "human" agent occurred by chance (see Chapter 1) rather than as a result of such specific attempts. Data concerning the disease potential of the newly isolated virus also suggest that the mechanism(s) governing the interaction of B19 virus and its human host are widely comparable to those revealed for the well-studied animal pathogens. Therefore, a concise discussion of the various aspects of parvovirus disease in animals might be helpful in sketching the pathogenic potential of this "new" human pathogen and in estimating its true role in human disease. In pursuing this goal, emphasis will be placed on general aspects such as the course of infection observed under natural and experimental conditions, the most common presentation of parvovirus disease, and last but not least, the present state of insight into the mechanisms underlying the pathogenicity of animal parvoviruses.

II. DIVERSITY, EPIDEMIOLOGY, AND VARIABILITY OF ANIMAL PARVOVIRUSES

First isolations in the late 1950s and in the early 1960s pointed to rodents — rats, hamsters, and mice — as the predominant natural hosts of parvoviruses.[1-2,11-16] Soon afterwards viral agents with comparable physicochemical properties were recovered from a wide variety of further animal species. The viruses were usually isolated from diseased animals by inoculation of tissue homogenates or fecal samples into cell cultures of animal and/or human origin.[12-13,16] However, what initially appeared to indicate a clear causative relationship between clinical illness and isolated pathogen required reconsideration after similar viruses were detected in "uninfected" cell cultures,[3-4] in media ingredients,[17] or as contaminants in adenovirus stock suspensions.[18] In such instances, screening of animal sera for antibodies to the isolated agents frequently helped to locate the respective natural host.

The actual state of our knowledge on the distribution of parvoviruses in animals and man is summarized in Table 1. Serological screening revealed that the great majority of these viruses are highly endemic in their natural host. For example, HI-antibodies to RV and to H-1 virus were found in rat colonies at an incidence of more than 80%,[19-20] antibodies to bovine parvovirus (BPV) were present in 86% of calf sera,[16-21] and 78% of pig sera were

Table 1
NATURAL HOST RANGE OF ANIMAL PARVOVIRUSES

Virus	Vernacular name	Natural host
RV	RAT virus	Rat
RTV		Rat(?)
TVX		Rat(?)
H-1		Rat, man(?)
HB		Man(?), hamster(?)
LuIII		Unknown
MVM	Minute virus of mice	Mice, rat
FPV	Feline parvovirus	Cats
Variants:		
MEV	Mink enteritis virus	Mink, raccoon
CPV	Canine parvovirus	Dogs
MVC	Minute virus of canines	Dogs
BPV	Bovine parvovirus	Cattle
PPV	Porcine parvovirus	Pigs
LPV	Lapine parvovirus	Rabbits
ADV	Aleutian disease virus	Mink
GPV	Goose parvovirus	Geese
Chicken PV	Chicken parvovirus	Chicken
HPLV[a]	Hepatopancreatic-parvo-like virus	Penaeid shrimp
B19		Man

[a] It is not yet clear whether this virus is a true member of the autonomous parvoviruses or whether it has to be classified with the Dependoviruses or the Densoviruses (insect parvoviruses).

positive for anti-PPV antibodies.[22] This situation is clearly favored by the tendency of the agents to establish latent, clinically inapparent infections, the shedding of viruses by such infected animals in feces, urine, saliva, and nasal secretions,[23-34] and, last but not least, by the outstanding stability of the virus particle in the environment. It proved to be almost impossible to prevent susceptible animals from contracting infection if, though housed in separate cages, they were kept in quarters with animals of the same species which carried the virus. In addition, the stability of the virus particle guarantees its prolonged survival in a contaminated environment even under the most unfavorable conditions. Canine parvovirus (CPV) was shown to survive in feces at room temperature for as long as 6 months[33] and Johnson and co-workers[35] reported that susceptible pigs picked up PPV almost immediately after being transferred to a stable formerly occupied by infected animals but routinely cleaned and kept empty for as long as 135 days.

With the exception of the host range variants of feline parvovirus — FPV, MEV, and CPV — all the viruses listed in Table 1 were shown to be antigenically distinct. FPV, MEV, and CPV can hardly be distinguished by routine serological techniques such as neutralization, hemagglutination-inhibition, or complement fixation with standard antisera. The use of monoclonal antibodies, however, seems to provide a useful tool to discriminate among the viruses.[36-37]

A number of possible parvoviruses have not been included in Table 1: a parvovirus-like agent has been detected in the feces of young turkeys suffering from diarrhea and enteropathy.[38] The agent has not yet been propagated in cell culture and no information concerning its antigenic relationship is available.

There is a presumed second parvovirus of cattle which was said to be antigenically distinct from bovine parvovirus (BPV).[39] This virus was isolated only once, its serologic relatedness to parvoviruses other than BPV obviously has never been tested, and there is circumstantial evidence suggesting that the isolate represented was in fact porcine parvovirus (PPV).[39a] The latter parvovirus is known to contaminate bovine cell cultures. A virus with all the physicochemical properties of a parvovirus has been isolated from extracts of the liver and lung of aborted equine fetuses via inoculation into pig fallopian tube cell cultures.[40] Recent serological studies seem to indicate that this virus, too, is antigenically closely related to if not identical with PPV.[40a] Because both primary and permanent porcine cell cultures are well known for frequently harboring latent PPV infections, the significance of the findings of Wong et al.[40] remains uncertain at the present time.

The situation encountered with the presumed second type of bovine parvovirus as well as with the agent recovered in the context of abortions in equines highlights a major problem in parvovirus research — the frequent contamination of cell cultures with these viruses. As will be outlined below in more detail, the latent presence of parvoviruses in primary cell cultures reflects directly the frequency of latent, persistent parvovirus infections in the animal host. Latent infections, both in vivo and in vitro, may go undetected for a long time. For example, RT virus, an antigenically distinct parvovirus of rats, has been isolated independently by two laboratories from a permanent rat cell line passaged around 70 and 120 times, respectively, after Tournier established the cells in vitro.[4,40b]

Parvoviruses can also be introduced unknowingly into cell cultures of heterologous origin. A long-term study dealing with these aspects has been reported by Hallauer et al.[4] These authors revealed the presence of parvoviruses in 38 out of 43 strains of permanent human cell lines (HeLa, KB, Hep-2, Detroit 30A, FL, Lu106, and Lu132) collected from a number of European laboratories. The vast majority of the isolated agents (29) proved to be serologically related to PPV. This virus was thought to have been introduced into the cells by use of contaminated trypsin, an assumption subsequently substantiated by the direct recovery of PPV from commercial trypsin of hog pancreas origin.[17] An additional eight of Hallauer's isolates represented a so far unknown serotype (TVX). Results of limited serological surveys point to the rat as the original host of this virus. How TVX had been introduced into the human cell lines remains a matter of speculation. Finally, even extended serologic screening of several thousand sera from a wide variety of species failed to yield any information concerning the natural host of LuIII virus. This agent is an antigenically distinct parvovirus and was recovered by Hallauer and co-workers[4] in but one single instance.

The possible presence of parvoviruses in cell cultures and the easy spread of the agents in the laboratory have several unpleasant consequences. Thus, whenever parvoviruses are isolated from clinical samples, the etiologic relationship between isolate and clinical disease has to be established beyond doubt. Moreover, reactivation of latent parvovirus infections may lead to contamination of harvests of other viruses and may interfere in an unpredictable manner with cell culture studies. Finally, latent viruses are likely to be picked up in vaccines and other cell culture-derived pharmaceutical products. These then may act as effective vehicles in the spread of the agents among susceptible animals. The recent sudden appearance and rapid spread of canine parvovirus in canine populations in all parts of the world may well be the result of such an event.[41-43]

The variants of feline parvovirus may also serve as an example to illustrate variation in host range among parvoviruses. As mentioned previously, FPV, MEV, and CPV cannot be distinguished routinely by conventional serological means. Infections with the feline parvovirus have been constantly recognized during the past hundred years. In the first half of this century, huge epidemics of feline enteritis have led to extensive experimental attempts to identify the etiologic agent.[23,44] Neither under field conditions nor upon experimental inoculation has it thereby been possible to transmit the virus to dogs or to other animals. In 1947, however, a disease similar to feline enteritis was recognized among ranch mink in Ontario.[45] It swept through Canada in serious epizootics between 1950 and 1952[46] and has now reached a world-wide distribution. Today, the causative agent — mink enteritis virus — cannot be distinguished from contemporary isolates of FPV. Like FPV, MEV is also not able to replicate and to induce disease in dogs.

What triggered the genetic changes enabling FPLV to extend its host range to mink will probably never be known. However, when in 1978 a virus with many properties of FPLV/MEV was found to be responsible for sudden and widespread epidemics of infectious enteritis in dogs, suitable experimental techniques were available to test some theories concerning the origin of the "new" pathogen. Comparative restriction site mapping of the genomes of CPV strains isolated in far apart geographical regions thus revealed that all infections — whether occurring in the U.S., Europe, or Australia — were caused by one single type of virus. Using the same technique, CPV could also be readily distinguished from isolates of FPV and MEV, and a significant relationship to two virus strains used in live attenuated FPV/MEV vaccines was established.[42,47] The studies provided no direct proof that the variant virus now known as CPV was derived from the vaccine viruses. However, considering the practice of attenuating viruses by deliberate and forced passage in heterologous cell culture systems and recalling the ease with which cross contamination with parvoviruses occurs in the laboratory, it does not seem to be too far fetched to assume that a variant of FPV/MEV able to replicate in canine cells originated in a laboratory. If such a virus gained access to a canine vaccine or some other product derived from canine cell cultures, the rapid spread of canine parvovirus in spite of all legal quarantine regulations can also be easily explained.

To date, FPV and its variants are the only parvoviruses known to cause severe clinical disease in more than one species. The ability to cross species barriers, however, has also been observed for other viruses listed in Table 1. Based on the incidence of antibodies, the natural host of H-1 virus seems to be the rat;[20] yet, the virus has repeatedly been isolated from tissues of human placentas and aborted fetuses via inoculation into newborn hamsters.[48-51] In related serologic surveys, anti-H-1 antibodies could also be demonstrated in a small percentage (usually about 1%) of human sera.[26,52] A markedly higher incidence of anti-H-1 antibodies (13.7%) was observed only once by Guglielmino et al.[53] in sera of women with a history of repeated abortions and stillbirth. To explain the difference in the incidence of H-1 infections in man as observed by different authors,[49,51,54] Guiglielmino and co-workers proposed that H-1 infection is a zoonosis, the incidence of positive human sera reflecting the likelihood of contact between man and rat. The opposite point of view, i.e., the rat acquiring H-1 infection from man, is held by Toolan,[54a] who points to the fact that sera with the highest incidence of anti-H-1 antibodies have been collected from rats feeding on human garbage dumps.[20]

A further parvovirus which evidently has the ability to replicate in cells of various species is porcine parvovirus. PPV is a frequent contaminant in the cell culture laboratory and, besides being found in cells of other species, easily manifests in cells of human origin. However, all attempts to demonstrate anti-PPV antibodies in sera of laboratory technicians, farmers, and other individuals in contact with the virus or with infected pigs have failed. In spite of these negative results, PPV as well as all the other members of the parvovirus genus have to be regarded as agents with a potential to establish themselves as pathogens in different host species. Therefore, experiments likely to yield new host range variants of parvoviruses have to be conducted with great care.

III. PATTERNS OF DISEASE

A. The Basis of Parvovirus Pathogenicity

The spectrum of parvovirus disease ranges from acute, lethal affections to slowly progressing, sometimes chronic illness. These presentations may occur in the course of wide spread epidemics or may be observed as completely isolated cases. Whether the clinical

Table 2
FACTORS INFLUENCING PARVOVIRUS PATHOGENICITY

Virus		Host	
Genetic background	**Function**	**Level**	**Factors**
Specificity	Cell species	Cell	Differentiation
Virulence	?	Tissue	Proliferation
		Organ	Regeneration
		Organism	Genetic background, immunity, copathogens
		Population	Epidemiologic situation

picture of a parvoviral infection develops in one or the other way is determined by an interplay of several factors and conditions. They are summarized in Table 2.

The most obvious of these factors are the genetic outfit of the virus and the genetic background of the host organism. This genetic information evidently determines the host range of a parvovirus and also favors the predominance of certain syndromes in a distinct virus/host system. The situation is exemplified best by the predominant manifestation of Aleutian disease in mink of the homozygous Aleutian genotype.[55] In addition, mutant viruses have been selected from strains of well-known parvoviruses which specifically replicate in only one type of cell and, hence, can be assumed to manifest in a rather distinct way in an infected organism.[56-61] Under natural conditions, however, there is now increasing evidence that, upon careful examination, the clinicopathological manifestations observed in one parvovirus/host system are likely to be found in every other virus/host system. Gradual differences in clinical presentation from system to system are most likely due to variations in the complexity and the time course of differentiation of host organisms.

As will be illustrated continuously during discussion of specific patterns of parvovirus disease, the main prerequisite for successful replication and manifestation of parvoviruses are rapidly proliferating cells or, more specifically, cells passing actively through the cell cycle. According to the results of a large series of experimental studies,[10,62] replication of parvoviruses in susceptible cells depends on cellular functions provided only shortly during S-phase and, most likely, towards the end of this phase of cellular DNA synthesis. Consequently, susceptible cells in differentiating tissues of embryos, fetuses, and neonates will provide excellent substrates for the replication of the viruses. Potent support for virus replication is also supplied by constantly or temporarily proliferating tissues in the adult organisms. On the other hand, parvovirus infections in susceptible but resting cells remain silent unless these cells are induced to proliferate. This can be achieved during regenerative growth in the course of wound healing, by tumorogenic transformation, or by co-occurring viral, bacterial, and parasitic infections.

The terms "susceptibility" and "differentiation" require further definition. "Susceptible" for parvoviruses means that an organism, a certain tissue, or a specific cell can be infected with and can support replication of the agent(s). The ability to support synthesis of progeny virus depends on cellular factors supplied exclusively by cycling cells. However, experimental observations, as well as a large body of circumstantial evidence, suggest that S-phase specific functions are but one requirement for this capability. The second, and probably most important requirement concerns "differentiation" of cells to a distinct, but in terms of physiological events not yet definable, developmental stage. Experimental examples for the existence and necessity of this developmental stage have been reported by Tattersall and co-workers.[56-58] Its importance for virus replication as well as for the expression of virus pathogenicity is illustrated best by the evident restriction of the replication of B19 virus to one (or very few) differentiational stages in erythrocyte development (see Chapter 8). Moreover, manifestation of perinatal parvovirus infections as distinct syndromes such as ataxia and myocarditis rather than as generalized disease might result from restriction of virus replication to tissues constituted by cells in an optimum state of differentiation.

An additional factor controlling replication, spread, and, hence, pathogenicity of parvoviruses at the level of the organism are the developmental stage and state of immunological competence. In this context, immunity in pregnant animals certainly plays an important role in preventing parvovirus infection in utero and, by passively transmitted antibody, also during the postpartum period. At the same time, the extent to which fetuses of, for example, pigs and bovines become competent to generate an immune response against an infecting parvovirus, seems to correlate with restriction of viral replication and spread.[63-66] In combination with a changing susceptibility of organs due to progressive differentiation of cells, this situation could help to limit virus synthesis and virus-specific damage to well-circumscribed areas of susceptible, yet to antiviral antibody only partly accessible, tissues.

Finally, the degree to which damage will become manifest in the form of a specific clinical syndrome is also likely to depend on the regenerative ability of the affected tissue. Cytolytic replication of parvoviruses in organs such as the liver, the regenerative ability of which remains at a high level in most animals for a considerable period of time during the postnatal period will hardly leave "scars" that become visible as distinct syndromes. Congenital malformations like cerebellar hypoplasia,[67-69] hydrocephalus,[70-71] and perhaps also myocarditis,[72] however, may well be the result of the target organ lacking regenerative potential.

A last factor with direct impact on the frequency of overt clinical disease induced by a parvovirus in populations of its natural host is herd immunity. Under conditions of high endemicity, overt disease is a rarity. Offspring born to an immune animal are likely to be protected by maternal, passively transmitted antibody. Under this protective passive immunity they will be continuously exposed to a parvovirus excreted by latently infected, clinically healthy animals and, in general, will develop active immunity. If, however, massive infection coincides with the decline of passive immunity, the full spectrum of clinical disease may ensue.

Populations of susceptible animals in which parvovirus infections show a morbidity of up to 100% are almost exclusively found in isolated locations[73] or under conditions where at least a certain degree of containment can be maintained, i.e., on well run, isolated animal ranches, or in efficiently shielded animal quarters. Yet, once a parvovirus has gained access to these places, an endemic situation will establish almost immediately. It can only be reversed by completely removing the infected animal stock and starting off with new, so far unexposed animals in a strictly disinfected, appropriately shielded environment.

An example illustrating both the rapidity of spread and the changing spectrum of clinical disease upon introduction of a parvovirus into a fully susceptible population is provided by the recent pandemic of CPV.[43,74] This previously unknown host range variant of feline parvovirus spread world-wide in the then virgin canine populations within only 1 to 2 years.[74] In canine populations hit by the virus for the first time, myocarditis in puppies was a relatively frequent finding[75] and enteritis caused severe losses in commercial kennels.[74] Nowadays, myocarditis is rarely if ever observed and cases of enteritis are mostly restricted to animals infected at the time of declining passive immunity or after inappropriate vaccination.[75-76]

A last group of animals susceptible to the attack of parvoviruses under an endemic situation consists in those, whose immune system has been disturbed by natural causes (malnutrition, microbial, and parasitic infection), or experimental and therapeutic treatment. In fact, the rodent parvovirus RV has been repeatedly isolated and disease due to RV could be frequently recorded when rats were either X-irradiated or treated with immunosuppressive chemicals.[11-13,77-78] It is not very clear whether the virus isolated and the clinical disease observed in these instances stemmed from *de novo* infection of sensitized animals or whether the various forms of treatment only conferred latent parvovirus infection into frank, clinical disease.

B. Disease Due to Infection In Utero

According to the basic considerations summarized in the foregoing chapter, several factors determine the probability, frequency, and extent of clinical manifestations in the fetus and the newborn if pregnant animals are exposed to parvoviruses:

1. Herd immunity controls the likelihood of acute progressive infection in the mature animal and, hence, the likelihood of transplacental infection.
2. Viral replication and manifestation of disease will only take place in tissues and/or organs undergoing differentiation and, hence, comprising a large proportion of actively dividing (''cycling'') cells.
3. The susceptibility of differentiating tissues and/or organs is determined by the developmental stage of the embryo or fetus; the state of differentiation of tissues and/or organs at the time of infection determines the regenerative potential and, hence, has a direct impact on the severity and persistence of induced disease.
4. The ability of the fetus and newborn organism to muster an immunological response to the infecting parvovirus can modify the extent of virus replication and may limit spread of infection within the organism.

These factors are not restricted to interaction of parvoviruses with the developing embryo and fetus. To a varying extent, they reach out into the postpartum period and affect replication of the agents in the newborn as well as the maturing animal. Syndromes known to result from in utero infection therefore will also be observed in one or the other form after infection of neonates, the immature organism, or even in the adult host.

1. Fetal Death and Congenital Malformations

Both experimental studies and circumstantial observations indicate that intrauterine infections with parvoviruses frequently result in death of the embryo and fetus as well as in obvious malformations.[10,26,79] Infections will usually occur by the transplacental route in nonimmune pregnant animals which suffer from acute parvovirus infection and develop a viremic state. However, the isolation of parvoviruses from scrotal lymphnodes, from testicles, and from semen of infected boars[31,80-82] also points to the possibility that the agents can be sexually transmitted. Although ova and the preimplantation stages of the early developing embryo appear to be refractory to infection with parvoviruses,[83-88] infection may proliferate in the vegetative cells of the ovary[20] and in the uterine epithelium and, later on, can reach and damage the developing fetus.[89]

Parvoviruses exert the most drastic effect on the developing fetus toward the end of the first and during the second trimester of gestation. During that time, hamster embryos infected with H-1 or LuIII virus or bovine and porcine embryos infected with BPV and PPV, respectively, mostly die and are absorbed or aborted.[26,64,66,86-88,90-91] In the pig, mummified fetuses have been observed to be farrowed at term. Littermates surviving to parturition may also be born severely ill, but even apparently normal animals then show a reduced vitality and an increased rate of neonatal death.

Infection after midgestation and especially in the third trimester are likely to result in distinct syndromes such as cerebellar hypoplasia,[67-69] hydrocephalus,[70] or myocarditis.[72] Moreover, infections occurring at this time may frequently be overcome without any clinical manifestation. Besides the changing spectrum of susceptible tissues in the differentiating organism, the main factor contributing to this resistance of the hamster, bovine, and porcine fetus to infection with the respective parvoviral pathogen probably resides in the then rapidly developing ability of the fetuses to generate an immune response.

Comparable to in utero infections in mammals, infection of chicken embryos with the recently isolated chicken parvovirus led to significantly reduced egg hatchability. In a study reported by Kisary[92,93] only 30 to 40% infected embryos hatched. They showed serious retardation in growth, bad feathering, as well as bone disorders and, frequently, suffered from massive enteritis. A total of 80 to 90% of such chicks died within a week of hatching.

2. The Osteolytic Syndrome

The most dramatic effect of perinatal parvovirus infection occurs in the induction of the so called "osteolytic syndrome". This syndrome has been extensively studied in hamsters infected either in utero or early after parturition with RV and related isolates,[44] the H-viruses,[26] as well as with MVM.[15] Characteristically, affected animals develop mongoloid-like features with small flat face, microcephalic domed head, protruding eyes and tongue, missing or abnormal teeth, and fragile bones.[2,95-100] These changes obviously result from the cytolytic attack of parvoviruses on osteogenic tissues such as the costochondral junctions of long bones,[95] the differentiating cranial sutures,[98] or the developing denture.[96-97,99,101-104] In every instance, virus replication leads to an improper and incomplete differentiation of the affected structure resulting in fragile bones, microcephalic domed head, and abnormalities in the denture varying from limited affections of individual teeth to complete destruction of the denture. (For a detailed summary of related pathology, see Siegl.[10]) In all instances, replication of parvoviruses at the location of lesions could be detected by the presence of typical intranuclear inclusion bodies. In hamsters, rats, and mice the attack of parvoviruses on the developing skeleton results in dwarfism of the affected animal. Therefore, the skeleton of the animals in general is affected in all its parts and the proportionality usually is conserved. Dwarfism and "stunted growth" have also been reported to occur subsequent to infection of calves with BPV, of piglets with PPV, of goslings with GPV, and of chickens with the newly discovered chicken parvovirus.[25,31,92-93,105] In these latter species, the differences in size between affected animals and uninfected controls is frequently overcome with progressive age. In hamsters and rats, however, the picture of dwarfism persists even in the aged animal.

"Mongoloid" hamsters and rats and probably all animals surviving perinatal parvovirus infection carry the agent throughout their life.[26] Such animals also show high titer antibodies to the infecting parvovirus in their serum.

3. Cerebellar Ataxia

About a century ago, Herringham and Andrews[106] described the clinical and pathological features of cerebellar ataxia in kittens. This syndrome was regarded as being of genetic origin until Kilham and Margolis[107] were able to induce classical ataxia in hamsters by injecting a distinct strain of RV into animals less than 4 days of age. Cerebellar ataxia was also observed in rats after transplacental infection with RV[69] and, finally, Kilham and Margolis,[68] and Johnson and co-workers[108] succeeded in inducing cerebellar ataxia with FPV in newborn kittens and in the ferret. According to a large body of experimental data[109,110] and circumstantial evidence, the vast majority of spontaneous cases of ataxia in kittens results in fact from in utero infections.

Histological examinations of the brains of ataxic hamsters, rats, ferrets, and kittens revealed the pathologic lesions accompanying clinical ataxia. Upon sacrifice of affected animals, the most prominent pathological finding always consisted of a severe hypoplasia of the cerebellum. It appeared to be due to the selective destruction of the dividing cells in the outer germinal layer of the differentiating cerebellum. The presence of basophilic intranuclear inclusion bodies signaled the attack of the virus upon this structure, and the degree of ensuing clinical disease evidently depended on the stage of differentiation of the definitive granular cortex at the time of infection. Maximal destruction of the germinal layer resulted from infection in utero and neither pathological nor clinical signs were observed when a fully differentiated cerebellum in a maturing animal was exposed to the virus.

Clinical ataxia has so far only been reported for infections with RV and FPV. Brain lesions resembling those induced by the cited viruses, however, were also observed in susceptible bovine fetuses infected with BPV at 130 days of fetal development.[110a] That clinical ataxia has not been described more frequently in animals affected by parvoviruses is most likely due to the high prevalence of these agents and hence the high incidence of immunity in their natural hosts. Upon introduction of the viruses into fully susceptible populations of host animals, however, cases of ataxia may be observed much more frequently.

4. *Myocarditis*

Myocarditis in puppies is another syndrome which apparently results from infection in utero the frequency of which seems to be controlled by the prevalence of the etiologic agent (CPV). At the time of the rapid spread of CPV in immunologically virgin canine populations during the late 1970s, myocarditis in 2- to 16-week-old puppies was a rather frequent finding.[75] The disease then presented itself in two forms. In puppies 2 to 8 weeks of age it was characterized by sudden death of the animals, occasionally preceded by a rather short period of dullness and dyspnea. In animals between 8 and 16 weeks of age, acute severe respiratory distress was observed about 24 hr before collapse and death occurred.[111-114] The mortality rate in affected litters varied from 20 to 100%. Survivors occasionally developed congestive heart failure. According to the very detailed histologic examinations of field cases (for references see Siegl,[79] and Meunier et al.[114]) the pathology of the disease presented as a marked interstitial myocarditis with loss of myofibers, multifocal myofiber necrosis, and, less constant, mononuclear cell infiltrate. In younger puppies, intranuclear inclusions could be demonstrated in myocytes at the time of necropsy.

Attempts to reproduce the disease under experimental conditions were only successful when pups were infected either in utero 8 days before[72] or in the early neonatal period up to 5 days after parturition.[114] Because a considerable proportion of myocardial cells in the dog undergo division in the perinatal phase it has been suggested[113] that, as is the case with cerebellar ataxia, this phase of rapid cell division provides the necessary substrate for manifestation of the virus in the myocardium of the very young susceptible canine. When, with increasing prevalence of CPV infections in canine populations and the continued use of vaccination, nonimmune pregnant bitches became a rarity, the frequency of parvoviral myocarditis dropped to almost zero.[75]

Widespread replication of parvoviruses in the myocardium has also been reported to occur in newborn kittens infected with FPV,[70-71,115] in calves infected with BPV,[30] as well as in goslings succumbing to infection with GPV.[25] None of these animals had manifest clinical myocarditis.

5. *Hepatitis*

Replication of rodent parvoviruses in the liver has been frequently observed in the course of in utero infections.[10,116] Due to the extended mitotic activity of hepatocytes in the postnatal phase of life and due to its outstandingly high regenerative potential, the liver also presents a prominent target for parvoviruses in the newborn, in the maturing, and even in the adult animal. Clinical hepatitis resulting from parvovirus infection therefore will be discussed at a later time in this article (see Section D.2).

C. Disease in the Newborn Animal

As has already been pointed out (see Section B.2), infection of an animal during the perinatal phase of life — whether in utero or shortly after parturition — results in largely identical patterns of disease. Depending on the developmental stage of the fetus or newborn animal, the clinical picture thereby may vary from generalized disease to the distinct syndromes described in the foregoing chapter. With progressive age, the termination of differentiation of tissues and the continuous maturation of the immune system increasingly favor the development of isolated syndromes unless the pattern of disease characteristic for the adult animal develops.

The one facet of parvovirus infection which seems to be characteristic for postnatal exposure is generalized disease with systemic involvement of almost all organs and tissues. Such generalized and often lethal infections have been frequently observed with rodent parvoviruses and with LuIII virus.[10] A large proportion of hamsters and rats infected with these agents early after birth died within 5 to 14 days. Clinical disease thereby comprised a rather short period of uncharacteristic symptoms like apathetic behavior, a sluggishness, and dyspnea. Acute generalized infection with a high rate of mortality was also found to be characteristic for infection of 4- to 8-day-old goslings with GPV and, though not yet studied in great detail, might result from infection of broiler chicken with the chicken parvovirus.[25,92-93,116] The clinical symptoms in geese and chicken comprise sluggishness, incoordination of movement, diarrhea, and nasal catarrh. In larger mammals such as cats and dogs, generalized disease in the newborn offspring appears to be a rare exception.[70-72]

D. Disease in the Maturing and Adult Animal

In a maturing animal, the number of target tissues susceptible to replication of parvoviruses and for manifestation of disease decreases to a state where only a few tissues with continuously elevated mitotic activity provide a substrate for virus synthesis. In the fully differentiated organism, such tissues are the intestinal epithelium as well as the hematopoietic/lymphopoietic system. Both tissue complexes are well known to respond to parvoviral infections. Besides, susceptible cells become available in episodes stimulating proliferation of distinct tissues for a limited period of time. Among such episodes one has to count pregnancy. Here the developing embryo or fetus represents the main target of viral attack, yet, maternal tissues such as the placenta will also support parvovirus replication. In fact, replication of parvoviruses in the placenta is thought to be a necessary prerequisite for infection of the fetus.[118] Moreover, regenerative growth in the course of wound healing,[119,120] cell proliferation stimulated by the presence of parasites,[121,122] hormonal and/or dietary stress,[122,123] or rapidly proliferating tumors[2,77] were found to generate the specific cell substrate for replication of the viruses.

Finally, parvoviral disease in the maturing and adult animal has to be discussed under the heading of direct horizontal infection from animal to animal as a consequence of restimulation of a preexisting, persistent, yet clinically unapparent infection. In both instances, the extent of clinical disease may vary considerably from mild, almost unrecognizable illness to one of the distinctly marked individual syndromes described below and, lastly, to severe disease comprising a complete spectrum of such syndromes. Moreover, whenever parvovirus infections take place, all susceptible tissues become involved and contribute to the clinicopathological picture of the resulting disease. Separate treatment of individual syndromes like enteritis and panleukopenia was chosen for the sake of clarity of presentation in the following paragraphs. This procedure, however, should not lead to the misconception that each syndrome, pronounced as it may be under distinct circumstances, will always occur as a separate entity.

1. Enteritis

Due to its epidemic appearance, frequently severe symptoms, as well as the sometimes extraordinarily high rate of mortality associated with it, enteritis is probably the best known individual syndrome connected with parvovirus infection. The spectrum of clinical signs, therefore, may be rather broad. It varies from barely noticeable disease consisting mostly of listlessness of the affected animal with borderline enteric signs, to heavy discharge of watery, sometimes blood stained stools in connection with pyrexia, anorexia, depression, dehydration, and not infrequently, death. Vomiting may be present, yet, is a less consistent feature.

Enteritis in its various forms has been observed in hamsters and rats,[10,26] cats,[124] dogs,[74] raccoons,[43] calves,[125] goslings,[25] chickens,[93] and, very likely, also in young turkeys.[38] In certain epidemics, diarrhea may represent the most prominent, most regularly occurring clinical symptom of parvovirus infections. Under controlled laboratory conditions, however, deliberate reproduction of the full picture of enteritis always proved to be difficult. Nevertheless, both successful and failing experiments yielded a concise picture of the prerequisites necessary for and the events leading to overt enteritis.

Under natural as well as under appropriate laboratory conditions, infection of a fully susceptible host with the respective parvovirus usually leads to the appearance of enteritis within 4 to 7 days. At that time, the affected animal may show the complete spectrum of clinical disease described before and, in severe cases, may die within 24 to 48 hr of onset of illness. During epidemics, the incidence of mortality in a fully susceptible population may be as high as 80%. In uncomplicated cases, on the other hand, animals usually recover rapidly after 3 to 4 days and the course of disease rarely exceeds 1 week.[43,74,79]

It is still under dispute whether the ingested virus replicates directly in cells of the intestinal tract,[122] or whether parvovirus infection by the oral route always leads to primary replication of the virus in the tonsils,[122,126-127] the thymus,[128] and other lymphoid tissues before the virus spreads via viremia and manifests in the epithelia of the intestine. Anyway, viremia usually can be observed on days 2 to 4 after infection concomitantly with the beginning of manifestations of infection in the crypt epithelium of the intestine.[70,71,122,125-135] In all species, the mucosa of the duodenum and the jejunum were found to be mainly affected, but the virus could also be detected in cells of the ileum, the colon, and the caecum. Cytolytic virus replication in these locations leads to necrosis of the crypt epithelium, atrophy of villi and, occasionally, to complete collapse of mucosal architecture.

Virus replication is usually at its height between 4 and 7 days post-infection when virus particles are being shed in the feces in huge quantities. Excretion of virus ceases sharply thereafter. Low, continuously diminishing concentrations of particles may nevertheless be present in stools of recovering animals. Recovery from disease is always signaled by the appearance of antibodies to the infecting virus as early as 3 to 5 days after infection.[74,126,129-131]

Actively dividing cells as the necessary substrate for the replication of parvoviruses are present in the intestinal epithelia of both young and adult, conventionally raised and specific inducible by natural and experimental infections, however, may vary tremendously. Recent experiments performed predominantly with cats, dogs, and calves now indicate that the severity of illness is linked to certain factors and conditions which in turn are directly related to the degree of mitotic activity in the intestinal mucosa. One of the factors affecting mitotic activity of epithelial cells evidently is the age of the animal. In the dog, for instance, cell proliferation in the intestine is low in suckling animals.[136] During weaning, which typically takes place at 5 to 7 weeks of age, the change of diet and the associated drastic change of the intestinal flora result in considerably increased mitotic activity. This observation may be relevant for the frequent failure in the past to induce canine parvovirus enteritis in very young, unweaned pups,[129] and may explain the severity and high incidence of enteritis usually occurring in weaned pups between 7 and 10 weeks of age.[111,130] In the mature and old animal, a reduced cellular proliferation in the intestine can be stimulated by dietary stress produced by starving and subsequent refeeding,[123] or by co-occurring viral, bacterial, and parasitic infections.[117,134,135,137] All the factors and conditions can operate in a synergistic way. Thus, Durham and co-workers[138] could show that infections with parvoviruses in 19-week-old freshly weaned calves led to little clinical response if the animals had no or only minimal coccidiosis. Disease, however, exacerbated substantially when animals were heavily infested with coccidia. The same authors also provided evidence that in animals carrying coccidia and undergoing infection with BPV before weaning, the stress of weaning evidently led to reactivation of subclinical parvovirus infection in the intestinal tract. The resulting postweaning diarrhea exceeded in severity by far the intestinal upset commonly associated with weaning of uninfected animals.

2. Hepatitis and Pancreatitis

In contrast to the situation in the intestinal mucosa, where mitotic activity is usually low during early postnatal life, differentiation of the liver and hence continued proliferation of hepatocytes occurs throughout a prolonged period of time from an early intrauterine state to definite maturation of the animal. Consequently, the liver provides a potent natural substrate for the replication of parvoviruses in the fetus, the newborn, and the premature animal.

In mammals, hepatitis related to infection with parvoviruses has mostly been observed in newborn rats and hamsters after spontaneous or experimental infection with RV.[69,139-141] Some of the affected animals were jaundiced and all of them showed histopathologic changes resulting from replication of parvoviruses in the liver. The changes consisted either in focal hepatocellular necrosis[140] or were described as cryptic lesions of the parenchymal peliosis type.[139,141]

The liver of the mature animal may be regarded as being refractory to parvovirus infection under normal circumstances. Nevertheless, Henry and Diorio[142] observed focal degeneration of hepatic cells and recorded elevated liver enzymes subsequent to infection of adult hamsters. Moreover, as in the case of the intestinal mucosa, regenerative growth subsequent to hepatic injury[116,143] or the presence of parasites in the liver[121] were shown to prime the liver of adult animals for manifestation of parvovirus infection.

Hepatitis due to infection with parvoviruses has also been observed in geese and most recently in penaeid shrimp. In goslings, hepatitis is one of the most prominent symptoms resulting from infection with GPV.[25,144] The parvovirus-like virus responsible for necrosis and atrophy of the hepatopancreas in shrimp, however, has not yet been fully characterized and typed.[145] Both viruses appear to be extraordinarily pathogenic because losses of up to 95% were reported to occur in affected geese flocks[117,146-148] and between 50 and 100% of animals in affected shrimp populations died within 4 to 8 weeks of onset of disease.[145]

Evidence for the involvement of parvoviruses in pancreatitis is scarce. Von Sandersleben and co-workers[149] observed insulin deficiency diabetes in four kittens 10 to 12 days of age. Upon necropsy, light microscopic and electron microscopic histology provided evidence for the destruction of exocrine pancreatic cells as well as of B cells by a parvovirus. The exact nature of the agent, however, could not be established.

3. Infection of the Central Nervous System

A complex of neurological symptoms other than cerebellar ataxia but nevertheless associated with parvovirus infection was recognized by chance by El Dadah et al., in 1967.[78] These investigators were able to isolate a strain of RV from the brain and spinal cord of rats which, after immunosuppressive treatment with cyclophosphamide, had developed paralysis of the hind limbs. In subsequent experiments, the virus proved to be just one out of several RV strains able to induce hemorrhagic encephalopathy — hemorrhages and necrosis in the brain and spinal cord associated with acute fatal paralysis — upon intracerebral injection into newborn rats, mice, and hamsters.[150-154] A second virus able to induce hemorrhagic encephalopathy in cats and ferrets proved to be feline panleukopenia virus.[154] Its host range variant CPV has recently been reported to cause severe necrotizing vasculitis and encephalomalacia in the course of disseminated infection in the dog.[155]

On the basis of the distribution of intranuclear, virus specific inclusion bodies in the various tissues of affected rodents, cats, and ferrets, Margolis and Kilham[116,154] hypothesized that the characteristic clinicopathologic features of the disease resulted from the cytolytic attack of the parvoviruses onto the proliferating vascular beds within the developing central nervous system as well as from cytolysis of developing megakaryocytes.

4. Infection of the Lymphopoietic/Hematopoietic System

If enteritis and diarrhea can be said to represent the most frequent, most dramatic facet of epidemically occurring parvovirus disease, the attack of parvoviruses on the lymphopoietic/hematopoietic system certainly has to be regarded as the most constant expression of parvovirus pathogenicity. This feature of parvovirus pathogenicity was first recognized in studies with "feline enteritis" in the years around 1940.[23,156-159] About three decades later, experimental studies with the same parvovirus then revealed that, although enteritis could be induced in cats only irregularly under the experimental conditions chosen, leukopenia could always be recorded.[73,133,137]

The feline panleukopenia virus/cat system now presents the best studied model for the attack of parvoviruses upon hematopoietic tissues. Comparable, if not basically indistinguishable results, however, have also been obtained with the other host range variants of feline parvovirus — MEV and CPV — in the respective hosts.[129,130,160,161] Of the other viruses listed in Table 1, pigs were found to regularly experience a transient leukopenia after initial exposure to PPV[162-165] and cytolytic infections in lymphatic tissues with BPV were recorded both in the developing fetus and in parallel to diarrhea in calves.[66,122,166] Finally, histopathologic observations recording changes in the morphology of lymph nodes and the thymus of rodents seem to indicate that in these species, too, the lymphopoietic tissues are constantly involved in the process of disease.[140] Hematologic changes, however, have not yet been reported. As indicated by the studies in cats, dogs, and mink, this failure may be explained by the early appearance, frequently rather brief duration, and mostly mild features of hematologic changes. The impact of parvovirus infections on the blood picture therefore may go undetected unless it is specifically looked for in well-organized follow-up studies, as has been done in the examples cited.

The well-documented follow-up studies with feline parvovirus and its variant viruses, as well as the experimental observations with PPV and BPV, indicate that these viruses replicate in tonsils, thymus, regional lymph nodes, and Peyer's patches as well as in the spleen within 24 to 48 hr after infection. Toward the end of this period and in parallel with a more or less pronounced viremia, the presence of the viruses in the bone marrow becomes evident by the occurrence of intranuclear inclusion bodies in the primitive blood cells. In infections with FPV the number of cells containing inclusion bodies within the bone marrow is usually small and their presence is usually short lived. In the case of Aleutian mink disease virus, however, up to 40% of marrow cells were found to contain specific intranuclear inclusions over a period of 15 months.[167] Cytolytic replication of the infecting parvovirus appears to result in quantitative alterations in the myeloid, erythroid, and megakaryocyte cell lines and, especially, in a neutropenia which is frequently severe. In a study reported by Boosinger et al.,[168a] however, infection with CPV failed to interfere with the normal maturation sequence of the predominant erythroid series.

The result of virus replication in the bone marrow as well as in the various lymphatic tissues of the infected animal becomes evident as more or less pronounced lymphopenia and panleukopenia at 2 to 5 days post-infection. In severe cases, the fall in white blood cell counts may be rather dramatic, with as few as 100 leukocytes per mm^3 being demonstrable at the height of disease 4 to 7 days post-infection. In such instances, prognosis is unfavorable and death may occur within a few hours. In general, however, leukopenia lasts for a period of less than 1 week. It is frequently followed by an episode of reactive leukocytosis and lymphocytosis.[122]

At the level of the lymphatic tissues, virus replication leads to lymphocytolysis in the spleen, the thymic cortex, and in the germinal centers of the lymph nodes. In dogs infected with CPV,[123,131] such alterations became evident as early as 2 to 3 days after infection. Between 5 and 7 days post-infection, the tissues showed severe depletion of thymocytes or lymphocytes, respectively. Repopulation of tissues occurred rapidly thereafter and, in uncomplicated cases, was completed by about 2 weeks.

In contrast to the situation with the human parvovirus B19 which is the main topic of this volume, a clear-cut effect of animal parvoviruses on erythropoiesis has so far never been reported, despite the fact that intranuclear inclusions have been repeatedly observed in the primitive blood cells within the bone marrow. It may well be that this facet of parvovirus pathogenicity escaped the notice of the various investigators in the field of veterinary virology due to the lack of a genetically determined condition associated with short-lived erythrocytes. However, one could also speculate that all animal parvoviruses known to date lack the ability to infect and/or to replicate in any of the erythrocyte progenitor cells.

The only effect on erythrocytes which repeatedly has been noticed during parvovirus infections in animals is an increased erythrophagocytosis.[128,137,168] According to O'Sullivan and co-workers[128] the reason for this might be seen in massive adsorption of the viral antigen onto the surface of erythrocytes.

E. Parvoviruses and the Immune System

It has been repeatedly stressed by several investigators that parvovirus infection and its effect onto the immune system of the infected host presents an enigma. On the one hand, the infecting virus replicates rapidly in the various lymphoid tissues and evidently causes a drastic decrease in all those cells involved in immune surveillance. On the other hand, antiviral antibody can already be found at about the time of the height of clinical disease with the increase in antibody titer evidently signaling and paralleling recovery of the host from infection. To complicate our understanding of the interaction of parvoviruses with the immune system, it is noted that the presence of an impressive humoral immunity does not necessarily equate with clearance of the virus from the host; rather, it appears to indicate or even depend on the continued latent presence of parvoviruses. Discussion of the interaction of parvoviruses with the immune system may finally be extended to Aleutian disease of mink, a slowly progressing, frequently fatal parvovirus infection in mink which is basically characterized by persistent synthesis of viral antigen and continued hypergammaglobulinemia.

1. Normal Immune Response to Parvovirus Infections

According to all available evidence, parvoviruses are excellent antigens. They are usually produced in large quantities during replication in vivo and stimulate an impressive humoral immune response in the immunocompetent animal host. In general, humoral antibody can be detected at the height of clinical disease, in cats and dogs for instance as soon as 4 to 7 days postinfection. In experimental CPV infections antibody titers have even been recorded as early as 3 days post-infection.[129-130]

Most likely, the antiparvovirus antibody persists at high titers throughout an animal's life. This suggests that the endemic presence of parvoviruses in populations of their natural hosts may favor repeated booster infections. However, there is also evidence that the usually high antibody titers observed in animals recovering from overt disease are directly related to the continued latent presence of the virus in the host. Proof for this situation may be taken from virus- and antibody-positive "mongoloid" hamsters,[26] from the frequent recovery of PPV out of the tissues of pigs with outstandingly elevated antibody titers,[22,80] or from the frequent reappearance of parvoviruses in association with clinical illness in animals subjected to immunosuppressive treatment.[11,13,78] In this context, Aleutian disease of mink where viral antigen and antiviral antibody are simultaneously and continuously produced after a prolonged period of time appears to be an extreme example. It will be discussed in more detail below.

Information concerning where parvoviruses may persist in a latent state is scarce. Paul and co-workers[168] found PPV to persist in the cytoplasm of peripheral blood B, T, and null cells. After these lymphoid cells were stimulated to proliferate by treatment with mitogens virus replication was again induced. Comparable observations were recorded in a laboratory system.[168b] Here, LuIII virus survived in noncycling, chemically arrested cells over periods of days and weeks. Release of such cells from the proliferation block then resulted in reinduction of virus synthesis and, finally, in cytolysis. It may therefore be speculated that parvovirus infection can be maintained in organs and tissues of low mitotic activity. Its perpetuation could be achieved by means of occasional cell divisions and by direct spread from cell to cell. This situation would help the virus and also virus-infected cells to escape the immune surveillance mechanisms of the animals. Favored by episodes of increased mitotic activity as associated with pregnancy, regenerative growth during wound healing, or rapidly proliferating tumors, increased virus replication could boost and perpetuate the immune response from within.

Protection conferred to the individual by a preexisting humoral parvovirus antibody has been a matter of much discussion among veterinarians involved in vaccination of cats, dogs, mink, and more recently of pigs. In general, the notion is held that antibody titers obtained by hemagglutination inhibition (HI) equal to or in excess of 16 can be considered as reflecting protective immunity, especially as neutralizing antibody titers in those sera usually exceed HI-titers by a factor of 10 to 100.[169,170] Such statements, however, become extraordinarily unreliable as soon as HI-titers are determined under unstandardized conditions. According to a recent report,[171] HI-antibody titers to CPV in a single canine serum varied from 4 to 800 when determined by several independent diagnostic laboratories.

2. The Impact of Parvovirus Replication on the Immunologic Responsiveness of the Infected Animal

The effect of parvovirus replication upon the mechanisms involved in humoral and cellular immune responses were investigated both in vivo and in vitro. Bonnard and co-workers[59] reported that a virus antigenically related to MVM was present in the culture fluid of a subline of the murine lymphoma cell EL-4. In contrast to the then known strains of MVM this virus grew in lymphocytes and proved to be immunosuppressive for allogenic leukocyte cultures. In closely related expriments, Campell et al.[172,173] isolated a variant of RV which also replicated in rat lymphocytes, depressed lymphocyte viability, and made infected lymphocytes unresponsive to stimulation with concanavalin A, phytohemagglutinin, or allogenic lymphocytes. The latter virus has not been studied further. Additional experiments[60,61] revealed, however, that the lymphotropic variant of MVM inhibited a variety of T-cell functions such as antigen-induced proliferation, generation of cytolytic T-cell activity, and T-helper functions. Moreover, attempts were made to trace the differences in host cell specificity between standard MVM and its immunosuppressive variant at the level of nucleotide sequences.[174] The respective results favor the idea that variation in the untranscribed region near the 3′ end of the genome may be responsible for the differences in cell tropism observed.

Darrigrand and co-workers[175] recently tried to assess the effect of a standard strain of RV on the immune system of two inbred strains of rats in the intact organism. In both strains of animal, the antibody response to virus infection proved to be identical. Natural cytotoxicity, however, varied. Comparable to results reported previously by Oehler et al.,[176] cytotoxicity was increased in Brown Norway rats at 3 to 4 days after infection. In contrast,

Wistar Furth rats developed a distinct depression of natural killer cell cytotoxic activity for a period of about 10 days. These results are in line with data derived from experimental studies in parvovirus-infected cats, dogs, and mink. Schultz et al.[177] could show that in the course of feline panleukopenia, responses to B-cell mitogens as well as antibody responses to viral and foreign cellular antigens were never impaired. In fact, clinically mild infections may even stimulate humoral immune mechanisms.[178] Lymphocyte responsiveness to phytohemagglutinin and concanavalin A, however, were regularly depressed. Almost identical results were obtained in dogs infected with CPV.[179]

In mink infected with Aleutian disease virus (ADV), the extent of immunological abnormalities evidently varies with the type of clinical disease and, hence, with the genetic background of infected animals and the genetic outfit of the infecting virus (see below). Animals with progressive disease consistently develop an antiviral lymophocyte response, a depressed response to T-lymphocyte mitogens, hypergammaglobulinemia with excessive synthesis of ADV specific antibody, immune complex formation, and last but not least, an autoimmune response.[180-185] In addition, suppression of responsiveness to a variety of antigens has been reported.[186-190] Neither an antiviral lymphocyte response nor immune suppression with regard to other antigens, however, has been observed in mink with nonprogressive Aleutian disease.[184]

3. *Aleutian Disease of Mink*

Aleutian disease of mink was first described in mink homozygous for the recessive gene of the Aleutian coat color.[191] It presents a rather complex virological and clinicopathological entity, a thorough discussion of which would exceed by far the scope of this article. The interested reader is referred to the excellent reviews of Porter et al.[55] and Lodmell et al.[192] for detailed further information. Here, emphasis will be placed on the immunological aspects of Aleutian disease.

Replication of Aleutian disease virus (ADV) takes place after infection of fetal, newborn, and adult mink. Only in mink of the Aleutian genotype, however, is the development of slowly progressive, fatal disease regularly observed. In mink of other genotype, clinical disease may be of variable duration and severity and, not infrequently, may take a subclinical course. Virus strain-specific variation in pathogenicity evidently adds to the outcome of infection.[55,193-195]

Characteristically, Aleutian disease takes a chronic, slowly progressive course. Clinical disease is associated with persistent infection, chronic viremia, and an abnormal, maximal immune response as reflected by the continued rapid proliferation of antibody-producing cells and by hypergammaglobulinemia. There is well-documented evidence that predominant pathologic changes and lesions are the result of an aberrent immune mechanism rather than the direct cytolytic effect of virus replication itself. Infected fully susceptible animals developing the complete spectrum of disease may die any time between 2 and 24 months after infection.[192,196-199]

Most of the pathological changes observed in Aleutian disease such as glomerulonephritis, polyarteritis, plasmacytosis, and splenomegaly have been attributed to the deposition of immune aggregates and complement in tissues or the persitence in the circulation of high concentrations of immunoglobulins. Evidence for the role of an adverse immunological response in the generation of ADV specific lesions has also been reported by Cheema et al.[200] who succeeded in preventing the development of lesions by immunosuppressive treatment. In addition, immunization of mink with killed ADV prior to infection with virulent virus clearly favored exacerbation of disease.[201] In spite of this situation, the initial antibody response of mink to ADV infection is not unusual. IgM antibody appears between 6 to 8 days after infection and persists for up to 85 days. IgG antibodies can be detected in largely normal concentrations from 10 to 12 days onward. By about 1 month after infection, however, the animals start to develop a marked increase in serum gammaglobulin which may eventually reach a maximal level of 15-fold that are seen in normal mink. The majority of the increased immunoglobulin is IgG and consists of antibody to ADV. IgA may also be markedly elevated and may be present throughout the course of infection.[202]

Immune complexes consisting of such antibody and of virus particles were found to be infectious and experimental attempts to neutralize ADV infectivity with sera of persistently infected mink as well as with sera raised in other species usually failed.[180,202] Therefore, it has been generally assumed that antibodies produced in the course of ADV infection lack virus neutralizing activity and thus, are not able to remove the virus from the organism. In very recent experiments, however, Stolze and Kaaden[208] could show that neutralizing antibodies to purified ADV are produced in normal titers in both naturally and experimentally infected mink. According to these results, lack of ability to neutralize the virus is not a property of the antibody but, rather is brought about by cellular structures related to phosphatidylcholin and cholesterin which bind to hydrophobic epitopes of the virus particle and thereby interfere sterically with the binding of antibody.

Additional abnormalities in the immune response of infected mink in the course of Aleutian disease consists in the appearance of an autoimmune component. Thus, positive direct Coomb's reactions have been reported,[203,204] antinuclear antibody and nuclear antigen have been demonstrated,[205] and anti-DNA antibody has repeatedly been described.[183,206,207] The nature and extent of the contribution of these abnormalities to the clinicopathological features of Aleutian disease, however, is not very clear. Finally, ADV infection evidently interferes with the immune response of the organism to nonviral antigens as has been outlined previously.

REFERENCES

1. **Kilham, L. and Olivier, L. J.,** A latent virus of rats isolated in tissue culture, *Virology*, 7, 428, 1959.
2. **Toolan, H. W., Dolldorf, G., Barclay, M., Chandra, S., and Moore, A. E.,** An unidentified filtrable agent isolated from transplanted human tumors, *Proc. Nat. Acad. Sci. U.S.A.*, 46, 1256, 1968.
3. **Hallauer, C. and Kronauer, G.,** Nachweis eines nicht identifizierten Haemagglutinins in menschlichen Tumorzellstaemmen *Arch. Gesamte Virusforsch.*, 11, 754, 1962.
4. **Hallauer, C., Kronauer, G., and Siegl, G.,** Parvoviruses as contaminants of permanent human cell lines. I. Virus isolations from 1960—1970, *Arch. Gesamte Virusforsch.*, 35, 80, 1971.
5. **Margolis, G. and Kilham, L.,** Rat Virus, An Agent with an Affinity for the Dividing Cell, Monogr. No. 2, National Institute of Neurological Diseases and Blindness, Bethesda, Maryland, 1966, 361.
6. **Kilham, L., Margolis, G., and Colby, E. D.,** Cerebellar ataxia and its congenital transmission in cats by feline panleukopenia virus, *J. Am. Vet. Med. Assoc.*, 158, 888, 1971.
7. **Johnson, R. H.,** Isolation of a virus from a condition simulating feline panleukopenia in a leopard, *Vet. Rec.*, 76, 1008, 1964.
8. **Johnson, R. H.,** Feline panleukopenia, I. Identification of a virus associated with the syndrome, *Res. Vet. Sci.*, 6, 466, 1965.
9. **Margolis, G., Kilham, L., and Johnson, R. H.,** The parvoviruses and replicating cells: insights into the pathogenesis of cerebellar hypoplasia, *Prog. Neuropathol.*, 1, 168, 1971.
10. **Siegl, G.,** The Parvoviruses, in *Virology Monographs*, Vol. 15, Gaerd, S. and Hallauer, C., Eds., Springer-Verlag, New York, 1976.
11. **Lum, G. S. and Schreiner, A. W.,** Study of a virus isolated from a chloroleukemic Wistar rat, *Cancer Res.*, 23, 1742, 1963.
12. **Zhdanov, V. M. and Merekalova, Z. J.,** Isolation of a virus from connective tissue of carcinogen-treated rats, *Vopr. Virusol.*, 7, 339, 1962.
13. **Payne, F. E., Shellabarger, C. J., and Schmidt, R. W.,** A virus from mammary tissue of rats treated with x-Rays or methylcholanthrene, *Proc. Am. Assoc. Cancer Res.*, 4, 51, 1963.
14. **Kilham, L.,** Rat virus (RV) infections in hamsters, *Proc. Soc. Exp. Biol. Med.*, 106, 825, 1961.
15. **Kilham, L. and Margolis, G.,** Pathogenicity of minute virus of mice (MVM) for rats, mice, and hamsters, *Proc. Soc. Exp. Biol. Med.*, 133, 1447, 1968.

16. **Abinanti, F. R. and Warfield, M. S.,** Recovery of a hemadsorbing virus (HADEN) from gastrointestinal tract of calves, *Virology*, 14, 288, 1961.
17. **Croghan, D. L., Matchett, A., and Koski, T. A.,** Isolation of porcine parvovirus from commercial trypsin, *Appl. Microbiol.*, 26, 431, 1973.
18. **Crawford, L. V.,** A minute virus of mice, *Virology*, 29, 605, 1966.
19. **Robey, R. E., Woodman, D. R., and Hetrick, F. M.,** Studies on the natural infection of rats with the Kilham rat virus, *Am. J. Epidemiol.*, 88, 139, 1968.
20. **Kilham, L. and Margolis, G.,** Transplacental infection of rats and hamsters induced by oral and parenteral inoculations of H-1 and rat viruses (RV), *Teratology*, 2, 111, 1969.
21. **Storz, J., Bates, R. C., Warren, G. S., and Howard, T. H.,** Distribution of antibodies against bovine parvovirus 1 in cattle and other animal species, *Am. J. Vet. Res.*, 33, 269, 1972.
22. **Bachmann, P. A.,** Vorkommen und Verbreitung von Picodna (Parvo)-Virus beim Schwein, *Zentralbl. Veterinaermed. Reihe B*, 16, 341, 1969.
23. **Sylvester, J. T., Lawrence, J. S., Ackart, R. J., Adams, W. S., Ervin, D. M., Haskins, A. L., Sounders, R. H., Stringfellow, M. B., and Wetrick, R. M.,** The virus of infectious feline agranulocytosis, I. Characters of the virus: pathogenicity, *J. Exp. Med.*, 77, 41, 1943.
24. **Gorham, J. R., Leader, R. W., and Henson, J. B.,** The experimental transmission of a virus causing hypergammaglobulinemia in mink: sources and modes of infection, *J. Infect. Dis.*, 114, 1964.
25. **Derzsy, D.,** A viral disease of goslings. I. Epidemiological, clinical, pathological and aetiological studies, *Acta Vet. Acad. Sci. Hung.*, 17, 443, 1967.
26. **Toolan, H. W.,** The picodna viruses: H. RV, and AAV, *Int. Rev. Exp. Pathol.*, 6, 135, 1968.
27. **Parker, J. C., Cross, S. S., Collins, M. J., Jr., and Rowe, W. P.,** Minute virus of mice II. Prevalence, epidemiology and occurrence as a contaminant of transplanted tumors, *J. Natl. Cancer Inst.*, 45, 1970.
28. **Binn, L. N., Lazar, E. C., Eddy, G. A., and Kajima, M.,** Recovery and characterization of a minute virus of canines, *Infect. Immun.*, 1, 503, 1970.
29. **Lipton, H., Nathanson, N., and Hodous, J.,** Enteric transmission of paroviruses: pathogenesis of rat virus infection in adult rats, *Am. J. Epidemiol.*, 96, 443, 1972.
30. **Storz, J. and Leary, J. J.,** Bovine parvoviruses: epidemiology and host-virus relationships, in Viral Enteritis, *INSERM*, 90, 63, 1979.
31. **Leman, A. D., Glock, R. D., Mengeling, W. L., Penny, R. H. C., Schott, E., and Straw, B.,** Porcine parvovirus infection, in *Diseases of Swine*, Leman, A. D. et al., Eds., Iowa State Press, Ames, Iowa, 1981, 352.
32. **Woods, C. B., Pollock, R. V. H., and Carmichael, L. E.,** Canine parvoviral enteritis, *J. Am. Anim. Hosp. Assoc.*, 16, 171, 1980.
33. **Pollock, R. V. H.,** Experimental canine parvovirus infection in dogs, *Cornell Vet.*, 72, 103, 1982.
34. **Joo, H. S., Donaldson-Wood, C. R., and Johnson, R. H.,** Antibody to porcine, feline and rat parvoviruses in various animal species, *Res. Vet. Sci.*, 21, 112, 1976.
35. **Johnson, R. H., Donaldson-Wood, C. R., and Joo, H. J.,** Observations on the epidemiology of porcine parvovirus, *Aust. Vet. J.*, 52, 80, 1976.
36. **Parrish, C. R. and Carmichael, L. E.,** Antigenic structure and variation of canine parvovirus type 2, feline panleukopenia virus, and mink enteritis virus, *Virology*, 129, 401, 1983.
37. **Parrish, C. R., Carmichael, L. E., and Antczak, D. F.,** Antigenic relationships between canine parvovirus type 2, feline panleukopenia virus and mink enteritis virus using conventional antisera and monoclonal antibodies, *Arch. Virol.*, 72, 267, 1982.
38. **Trampel, D. W., Kinden, D. A., Solorzano, R. F., and Stogsdill, P. L.,** Parvovirus-like enteropathy in Missouri Turkeys, *Avian Dis.*, 27, 49, 1983.
39. **Inaba, Y., Omori, T., Kono, M., Ishii, S., and Matumoto, M.,** A new serotype of bovine parvovirus, *Jpn. J. Microbiol.*, 17, 85, 1973.

39a. **Johnson, R. H.,** personal communication.

40. **Wong, F. C., Spearman, J. G., Smolenski, M. A., and Loewen, P. C.,** Equine parvovirus: initial isolation and partial characterization, *Can. J. Comp. Med.*, 49, 50, 1985.

40a. **Molitor, T. W.,** personal communication.

40b. **Bachmann, P. A.,** personal communication.

41. **Johnson, R. H. and Spradbrow, P. B.,** Isolation from dogs with severe enteritis of a parvovirus related to feline panleukopenia virus, *Aust. Vet. J.*, 55, 151, 1979.
42. **Tratschin, J. D., McMaster, G. K., Kronauer, G., and Siegl, G.,** Canine parvovirus — relationship to wild-type and vaccine strains of feline panleukopenia virus and mink enteritis virus, *J. Gen. Virol.*, 61, 33, 1982.
43. **Siegl, G.,** Canine parvovirus: origin and significance of a "new" pathogen, in *The Parvoviruses*, Berns, K. I., Ed., Plenum Press, New York, 1984, chap. 9.
44. **Zschokke, E.,** Ueber coli-bacillaere Infectionen, *Schweiz. Arch. Tierheilkd.*, 42, 20, 1900.

45. **Schofield, F. W.,** Virus enteritis in mink, *N. Am. Vet.*, 30, 651, 1949.
46. **Wills, C. G.,** Notes on infectious enteritis of mink and its relationship to feline enteritis, *Can. J. Comp. Med.*, 16, 419, 1952.
47. **McMaster, G. K., Tratschin, J. D., and Siegl, G.,** Comparison of canine parvovirus with mink enteritis virus by restriction site mapping, *J. Virol.*, 38, 368, 1981.
48. **Toolan, H. W.,** A virus associated with transplantable tumors, *Bull. N.Y. Acad. Med.*, 37, 305, 1961.
49. **Toolan, H. W.,** Studies on a viral agent associated with human tissues, *Proc. Am. Assoc. Cancer Res.*, 3, 273, 1961.
50. **Toolan, H. W.,** Studies on H-viruses, *Proc. Am. Assoc. Cancer Res.*, 5, 64, 1964.
51. **Toolan, H. W., Buttle, G. A. H., and Kay, H. E. M.,** Isolation of the H-1 and H-3 viruses directly from human embryos, *Proc. Am. Assoc. Cancer Res.*, 3, 368, 1962.
52. **Monif, R. G., Sever, J. L., and Cochran, W. D.,** The H-1 and the RV viruses and pregnancy: serological study of certain groups of pregnant women, *J. Pediatr.*, 67, 253, 1965.
53. **Guglielmino, S., Tempera, G., Pappalardo, G., and Castro, A.,** H-1 and X14 parvovirus antibodies in women with abortions and still-births, *Acta Virol.*, 22, 426, 1978.
54. **Newman, S. J., McCallin, P. F., and Sever, J. L.,** Attempts to isolate H-1 virus from spontaneous human abortions: a negative report, *Teratology*, 3, 279, 1970.
54a. **Toolan, H. W.,** personal communication.
55. **Porter, D. D., Larsen, A. E., and Porter, H. G.,** Aleutian disease of mink, *Adv. Immunol.*, 29, 261, 1980.
56. **Tattersall, P. and Bratton,** Reciprocal productive and restrictive virus-cell interactions of immunosuppressive and prototype strains of minute virus of mice, *J. Virol.*, 46, 944, 1983.
57. **Spalholz, B. A. and Tattersall, P.,** Interaction of minute virus of mice with differentiated cells: strain-dependent target cell specificity is mediated by intracellular factors, *J. Virol.*, 46, 937, 1983.
58. **Tattersall, P.,** Susceptibility to minute virus of mice as a function of host-cell differentiation, in *Replication of Mammalian Parvoviruses*, Ward, D. C. and Tattersall, P., Eds., Cold Spring Harbor Press, Cold Spring Harbor, N.Y., 1978, 131.
59. **Bonnard, G. D., Manders, E. K., Campbell, D. A., Herberman, R. B., and Collins, M. J.,** Immunosuppressive activity of a subline of the mouse EL4 lymphoma. Evidence for minute virus of mice causing the inhibition, *J. Exp. Med.*, 143, 187, 1976.
60. **McMaster, G. K., Beard, P., Engers, H. D., and Hirt, B.,** Characterization of an immunosuppressive parvovirus related to the minute virus of mice, *J. Virol.*, 38, 317, 1981.
61. **Engers, H. D., Louis, J. A., Zubler, R. H., and Hirt, B.,** Inhibition of T-cell mediated functions by MVM (i), a parvovirus closely related to minute virus of mice, *J. Immunol.*, 127, 2280, 1981.
62. **Margolis, G., Kilham, L., and Johnson, R. H.,** The parvoviruses and replicating cells: insights into the pathogenesis of cerebellar hypoplasia, *Prog. Neuropathol.*, 1, 168, 1971.
63. **Redman, D. R., Bohl, E. H., and Ferguson, L. C.,** Porcine parvovirus; natural and experimental infections of the porcine fetus and prevalence in mature swine, *Infect. Immun.*, 10, 718, 1974.
64. **Bachmann, P. A., Sheffy, B. E., and Vaughan, J. T.,** Experimental in utero infection of fetal pigs with a porcine parvovirus. *Infect. Immun.*, 12, 455, 1975.
65. **Cutlip, R. C. and Mengeling, W. L.,** Pathogenesis of in utero infection of eight- and ten-week-old porcine features with porcine parvovirus, *Am. J. Vet. Res.*, 36, 1751, 1975.
66. **Storz, J., Young, S., Carroll, E. J., Bates, R. C., Bowen, R. A., and Keney, D. A.,** Parvovirus infection of the bovine fetus: distribution of infection, antibody response, and age-related susceptibility, *Am. Vet. Res.*, 39, 1099, 1978.
67. **Kilham, L. and Margolis, G.,** Cerebellar disease in cats induced by inoculation of rat virus, *Science*, 148, 244, 1965.
68. **Kilham, L. and Margolis, G.,** Viral etiology of spontaneous ataxia of cats, *Am. J. Pathol.*, 48, 991, 1966.
69. **Kilham, L. and Margolis, G.,** Spontaneous hepatitis and cerebellar hypoplasia in suckling rats due to congenital infection with rat virus, *Am. J. Pathol.*, 49, 457, 1966.
70. **Csiza, C. K., Scott, F. W., de Lahunta, A., and Gillespie, J. H.,** Pathogenesis of feline panleukopenia virus in susceptibile newborn kittens. I. Clinical signs, hematology, serology, and virology, *Infect. Immun.*, 3, 833, 1971.
71. **Csiza, C. K., de Lahunta, A., Scott, F. W., and Gillespie, J. H.,** Pathogenesis of feline panleukopenia virus in susceptible newborn kittens. II. Pathology and immunofluorescence, *Infect. Immun.*, 3, 838, 1971.
72. **Lenghaus, C., Studdert, M. J., and Finnie, J. W.,** Acute and chronic canine parvovirus myocarditis following intrauterine inoculation, *Aust. Vet. J.*, 56, 465, 1980.
73. **Johnson, R. H.,** Feline panleucopaenia, *Vet. Rec.*, 29, 338, 1969.
74. **Carmichael, L. E. and Binn, L. N.,** New enteric viruses in the dog, in *Advances in Veterinary Science and Comparative Medicine*, Vol. 25, Cornelius, C. E. and Simpson, C. F., Eds., Academic Press, New York, 1981, 1.

75. **Sabine, M., Feilen, C., Herbert, L., Walker Lomas, S., Wild, J., Love, D., and McGavin, D.,** Canine parvovirus — five years later, *Aust. Vet. Pract.*, 14, 24, 1984.
76. **Pollock, R. V. H. and Carmichael, L. E.,** Maternally derived immunity to canine parvovirus infection. Transfer, decline, and interference with vaccination, *J. Am. Vet. Med. Assoc.*, 180, 37, 1982.
77. **Spencer, H. J.,** Recovery of rat virus from a series of chemically induced rat leukemia, *Proc. Am. Assoc. Cancer Res.*, 8, 62, 1967.
78. **El Dadah, A. N., Nathanson, N., Smith, K. O., Squire, R. A., Santos, G. W., and Melby, E. C.,** Viral hemorrhagic encephalopathie of rats, *Science*, 156, 392, 1967.
79. **Siegl, G.,** Biology and pathogenicity of autonomous parvoviruses, in *The Parvoviruses*, Berns, K. I., Ed., Plenum Press, New York, 1984, 297.
80. **Cartwright, S. F. and Huck, R. A.,** Viruses isolated in association with herd infertility, abortions and still births in pigs, *Vet. Rec.*, 81, 196, 1967.
81. **Lucas, M. H., Cartwright, S. F., and Wrathall, A. E.,** Genital infection of pigs with porcine parvovirus, *J. Comp. Pathol.*, 84, 347, 1974.
82. **McAdaragh, J. P. and Anderson, G. A.,** Transmission of virus through boar semen, *Proc. 18th Am. Assoc. Vet. Lab. Diagn.*, 69, 1975.
83. **Mohanty, S. B. and Bachmann, P. A.,** Susceptibility of fertilized mouse eggs to minute virus of mice, *Infect. Immun.*, 9, 762, 1974.
84. **Wrathall, A. E. and Mengeling, W. L.,** Effect of porcine parvovirus on development of fertilized pig eggs in vitro, *Br. Vet. J.*, 135, 249, 1979.
85. **Ferm, V. H. and Kilham, L.,** Rat virus (RV) infection in fetal and pregnant hamsters, *Proc. Soc. Exp. Biol. Med.*, 112, 623, 1963.
86. **Ferm, V. H. and Kilham, L.,** Congenital anomalies induced in hamster embryos with H-1 virus, *Science*, 145, 510, 1964.
87. **Ferm, H. and Kilham, L.,** Histopathologic basis of the teratogenic effects of H-1 virus on hamster embryos, *J. Embryol. Exp. Morphol.*, 13, 151, 1965.
88. **Soike, K. F., Iatropoulis, M., and Siegl, G.,** Infection of newborn and fetal hamster induced by inoculation of LuIII parvovirus, *Arch. Virol.*, 51, 235, 1976.
89. **Wrathall, A. E. and Mengeling, W. L.,** Effect of transferring parvovirus-infected fertilized pig eggs into seronegative gilts, *Br. Vet. J.*, 135, 255, 1979.
90. **Joo, H. S, Donaldson-Wood, C. R., and Johnson, R. H.,** Rapid diagnostic techniques for detection of porcine parvovirus infection in mummified foetuses, *Aust. Vet. J.*, 52, 51, 1976.
91. **Mengeling, W. L., Cutlip, R. C., Wilson, R. A., and Parks, J. B.,** Fetal mumification associated with porcine parvovirus infection, *J. Am. Vet. Med. Assoc.*, 166, 993, 1975.
92. **Kisary, J., Nagy, B., and Bitay. Z.,** Presence of parvovirus in the intestine of chicken showing stunting syndrome, *Avian Pathol.*, 13, 339, 1984.
93. **Kisary, J.,** Experimental infection of chicken embryos and day-old chickens with parvovirus of chicken origin, *Avian Pathol.*, 14, 1, 1985.
94. **Kilham, L.,** Mongolism associated with rat virus (RV) infection in hamsters, *Virology*, 13, 141, 1961.
95. **Dalldorf, G.,** Viruses and human cancer, *Bull. N.Y. Acad. Med.*, 36, 795, 1960.
96. **Baer, P. N. and Kilham, L.,** Rat virus and peridontal disease. I. The peridontum in mongoloid hamsters, *Oral Surg. Oral Med. Oral Pathol.*, 15, 756, 1962.
97. **Baer, P. N. and Kilham, L.,** Rat virus and peridontal disease. II. Onset and effect of age and time of inoculation, *Oral Surg. Oral Med. Oral Pathol.*, 15, 1302, 1962.
98. **Ferm, V. H. and Kilham, L.,** Skeletal studies of virus-induced dwarfism, *Growth*, 29, 7, 1965.
99. **Baer, P. N. and Kilham, L.,** A comparison of the effects of four viruses on the periodontium of the Syrian hamster, *Periodontology*, 36, 127, 1965.
100. **Lipton, H. L. and Johnson, R. T.,** The pathogenesis of rat virus infections in the newborn hamster, *Lab. Invest.*, 27, 508, 1972.
101. **Baer, P. N. and Kilham, L.,** Rat virus and periodontal disease. III. The histopathology of early lesions in the first molar, *Oral Surg. Oral Med. Oral Pathol.*, 17, 116, 1964.
102. **Baer, P. N. and Kilham, L.,** Rat virus and peridontal disease. IV. The aged hamster, *Oral Surg. Oral Med. Oral Pathol.*, 18, 803, 1964.
103. **Baer, P. N. and Kilham, L.,** Dental defects in hamsters infected with minute virus of mice, *Oral Surg. Oral Med. Oral Pathol.*, 37, 385, 1974.
104. **Cohen, M. M. and Shklar, G.,** Deformitives of the cranofacial and dental complex in the H-1 virus modified hamster, *Oral Surg. Oral Med. Oral Pathol.*, 17, 533, 1964.
105. **Storz, J. and Bates, R. C.,** Parvovirus infections in calves, *J. Am. Vet. Med. Assoc.*, 163, 884, 1973.
106. **Herringham, W. P. and Andrews, F. W.,** Two cases of cerebellar disease in cats with staggering, *St. Barth's Hosp.*, 24, 112, 1888.
107. **Kilham, L. and Margolis, G.,** Cerebellar ataxia in hamsters inoculated with rat virus, *Science*, 143, 1047, 1964.

108. **Johnson, R. H., Margolis, G., and Kilham, L.,** Identity of feline ataxia virus with feline panleukopenia virus, *Nature (London)*, 214, 175, 1967.
109. **Duenwald, J. C., Holland, J. M., Gorham, J. R., and Ott, R. L.,** Feline panleukopenia: experimental cerebellar hypoplasia produced in neonatal ferrets with live virus vaccine, *Res. Vet. Sci.*, 12, 394, 1971.
110. **Kilham, L. and Margolis, G.,** Problems of human concern arising from animal models of intrauterine and neonatal infections due to viruses: a review. I. Introduction and virologic studies, *Prog. Med. Virol.*, 20, 113, 1975.
110a. **Storz, J.,** personal communication.
111. **McCandlish, I. A. P., Thompson, H., Cornwell, H. J., and Macartney, L.,** Canine parvovirus infection, *Vet. Rec.*, 107, 204, 1980.
112. **Mulvey, J. J., Bech-Nielsen, S., Haskins, M. E., Jezyk, P. F., Taylor, H. W., and Eugster, A. K.,** Myocarditis induced by parvoviral infection in weanling pups in the United States, *J. Am. Vet. Med. Assoc.*, 177, 695, 1980.
113. **Robinson, W. F., Wilcox, G. E., Flower, R. L. P., and Smith, J.,** Evidence for a parvovirus as the aetiologic agent in myocarditis of puppies, *Aust. Vet. J.*, 55, 294, 1979.
114. **Meunier, P. C., Cooper, B. J., Appel, M. J. G., and Slauson, D. O.,** Experimental viral myocarditis. Parvoviral infection of neonatal pups, *Vet. Pathol.*, 21, 509, 1984.
115. **Bestetti, G. and Zwahlen, R.,** Generalized parvovirus infection with inclusion-body myocarditis in two kittens, *J. Comp. Pathol.*, 95, 393, 1985.
116. **Margolis, G. and Kilham, L.,** Problems of human concern arising from animal models of intrauterine and neonatal infections due to viruses: a review. II. Pathologic studies, *Prog. Med. Virol.*, 20, 144, 1975.
117. **Schettler, C. H.,** Isolation of a highly pathogenic virus from geese with hepatitis, *Avian Dis.*, 15, 323, 1971.
118. **Mengeling, W. L., Paul, P. S., and Brown, T. T.,** Transplacental infection and embryonic death following maternal exposure to porcine parvovirus near the time of conception, *Arch. Virol.*, 65, 55, 1980.
119. **Baer, P. N., Garrington, G. E., and Kilham, L.,** Effect of age and H-1 virus on healing fractures in hamsters, *J. Gerontol.*, 26, 373, 1977.
120. **Engler, W. O., Baer, P. N., and Kilham, L.,** Effects of rat virus on healing osseous wounds, *Arch. Pathol.*, 82, 93, 1966.
121. **Kilham, L., Margolis, G., and Colby, E. D.,** Enhanced proliferation of H-1 virus in livers of rats infected with *Crysticercus faciolaris*, *J. Infect. Dis.*, 121, 648, 1970.
122. **Durham, P. J. K., Lax, A., and Johnson, R. H.,** Pathological and virological studies of experimental parvoviral enteritis in calves, *Res. Vet. Sci.*, 38, 209, 1985.
123. **Carman, S. and Povey, S.,** Successful experimental challenge of dogs with canine parvovirus-2, *Can. J. Comp. Med.*, 46, 33, 1982.
124. **Urbain, A.,** Contribution a l'etude de la gastro-enterite infectieuse des chats, *Ann. Inst. Pasteur.*, 51, 202, 1933.
125. **Storz, J., Leary, J. J., Carlson, J. H., and Bates, R. C.,** Parvoviruses associated with diarrhea in calves, *J. Am. Vet. Med. Assoc.*, 173, 624, 1978.
126. **Carman, P. S. and Povey, R. C.,** Pathogenesis of canine parvovirus-2 in dogs — histopathology and antigen identification in tissues, *Res. Vet. Sci.*, 38, 141, 1985.
127. **Kahn, D. E.,** Pathogenesis of feline panleukopenia, *J. Am. Vet. Med. Assoc.*, 173, 628, 1978.
128. **O'Sullivan, G., Durham, P. J. K., Smith, J. R., and Campbell, R. S. F.,** Experimentally induced severe canine parvoviral enteritis, *Aust. Vet. J.*, 61, 1, 1984.
129. **Carman, P. S. and Povey, R. C.,** Pathogenesis of canine parvovirus-2 in dogs — haematology, serology and virus recovery, *Res. Vet. Sci.*, 38, 134, 1985.
130. **Macartney, L., McCandlish, I. A. P., Thompson, H., and Cornwell, H. J. C.,** Canine parvovirus enteritis. 1. Clinical, haematological and pathological features of experimental infection, *Vet. Rec.*, 115, 201, 1984.
131. **Macartney, L., McCandlish, I. A. P., Thompson, H., Cornwell, H. J. C.,** Canine parvovirus enteritis. II. Pathogenesis, *Vet. Rec.*, 115, 453, 1984.
132. **Langheinrich, K. A. and Nielsen, S. W.,** Histopathology of feline panleukopenia: report of 65 cases, *J. Am. Vet. Med. Assoc.*, 158, 863, 1971.
133. **Larsen, S. A., Flagstad, A., and Aalbaek, B.,** Experimental feline panleukopenia in the conventional cat, *Vet. Pathol.*, 13, 216, 1976.
134. **Carlson, J. H., and Scott, F. W.,** Feline panleukopenia. II. The relationship of intestinal mucosal cell proliferation rates to viral infection and development of lesions, *Vet. Pathol.*, 14, 173, 1977.
135. **O'Reilly, K. J., and Whitaker, A. M.,** The development of feline cell lines for the growth of feline infectious enteritis (panleukopenia) virus, *J. Hyg.*, 67, 115, 1969.
136. **Koldovsky, O., Sunshine, P., and Kretchmer, N.,** Cellular migration of intestinal epithelia in suckling and weaned rats, *Nature (London)* 212, 1389, 1966.

137. **Carlson, J. H., Scott, F. W., and Duncan, J. R.,** Feline panleukopenia. Pathogenesis in germfree and specific pathogen free cats, *Vet. Pathol.*, 14, 79, 1977.
138. **Durham, P. J. K., Johnson, R. H., and Parker, R. J.,** Exacerbation of experimental parvoviral enteritis in calves by coccidia and weaning stress, *Res. Vet. Sci.*, 39, 16, 1985.
139. **Margolis, G., Kilham. L., and Ruffolo, P. R.,** Rat virus disease, an experimental model of neonatal hepatitis, *Exp. Mol. Pathol.*, 8, 1, 1968.
140. **Coleman, G. L., Jacoby, R. O., Bhatt, P. N., Smith, A. L., and Jonas, A. M.,** Naturally occurring lethal parvovirus infection of juvenile and young-adult rats, *Vet. Pathol.*, 20, 49, 1983.
141. **Bergs, V. V. and Scotti, T. M.,** Virus-induced peliosis hepatitis in rats, *Science*, 158, 377, 1967.
142. **Henry, C. J. and Diorio, A. F.,** Induction of hepatitis in adult Syrian hamsters by H-1 virus, *Proc. Soc. Exp. Biol. Med.*, 149, 23, 1975.
143. **Ruffolo, P. R., Margolis, G., and Kilham, L.,** The induction of hepatitis by prior partial hepatectomy in resistant adults rats injected with H-1 virus, *Am. J. Pathol.*, 49, 795, 1966.
144. **Palya, V. and Kisary, J.,** Histopathological studies on goslings infected experimentally with the wild-type and attenuated virus of Derzy's disease, *Mag. Allatorv. Lapja*, 1978/12, 830, 1978.
145. **Lightner, D. V. and Redman, R. M.,** A parvo-like virus disease of penaeid shrimp, *J. Invertebr. Pathol.*, 45, 47, 1985.
146. **Coudart, M., Fedida, M., Dannacher, G., and Peillon, M.,** The late form of parvovirus disease of gosling, *Rec. Med. Vet.*, 150, 899, 1974.
147. **Krauss, H.,** Eine verlustreiche Aufzuchtkrankheit bei Gänseküicken (Erste Mitteilung), *Berl. Müench. Tieräerztl. Wochenschr.*, 78, 372, 1965.
148. **Dannacher, G., Fouillet, X., Coudert, M., Fedida, M., and Petillon, M.,** Etiologie de la maladie a virus de l'oison: le virus β, *Rec. Med. Vet.*, 150, 49, 1974.
149. **von Sandersleben, J., Pospischil, A., and Kraft, W.,** Infektion des Pankreas mit Parvoviren bei der Katze, *Dtsch. Tieräerztl. Wochenschr.*, 90, 297, 1983.
150. **Cole, G. A. and Nathanson, N.,** Hemorrhagic encephalopathy induced by cyclophosphamide in suckling rats, *Cancer Chemother. Rep.*, 54, 211, 1970.
151. **Cole, G. A., Nathanson, N., and Rivert, H.,** Viral hemorrhagic encephalopathy of rats. II. Pathogenesis of central nervous system lesions, *Am. J. Epidemiol.*, 91, 339, 1970.
152. **Nathanson, N., Cole, G. A., Santos, G. W., Squire, R. A., and Smith, K. O.,** Viral hemorrhagic encephalopathy of rats. I. Isolation, identification, and properties of the HER strains of rat virus, *Am. J. Epidemiol.*, 91, 328, 1970.
153. **Margolis, G. and Kilham, L.,** Rat virus infection of megakaryocytes: a factor in hemorrhagic encephalopathy?, *Exp. Mol. Pathol.*, 16, 326, 1972.
154. **Margolis, G. and Kilham, L.,** Parvovirus infections, vascular epithelium and hemorrhagic encephalopathy, *Lab. Invest.*, 22, 478, 1970.
155. **Johnson, B. J. and Castro, A. E.,** Isolation of canine parvovirus from a dog brain with severe necrotizing vasculitis and encephalomalacia, *J. Am. Vet. Med. Assoc.*, 184, 1398, 1984.
156. **Hammon, W. D. and Enders, J. F.,** A virus disease of cats, principally characterized by aleucocytosin, enteric lesions and the presence of intranuclear inclusion bodies, *J. Exp. Med.*, 69, 327, 1939.
157. **Hammon, W. D. and Enders, J. F.,** Further studies on the blood and the hematopoietic tissues in malignant panleukopenia of cats, *J. Exp. Med.*, 70, 557, 1939.
158. **Kikuth, W., Gonnert, R., and Schweikert, M.,** Infektioese Aleukozytose der Katzen, *Zentralbl. Bakteriol. Parasitenkd. Infektionskr. Hyg. Abt. 1: Orig.*, 146, 1, 1940.
159. **Lawrence, J. S., Syverton, J. T., Shaw, Jr., J. S., and Smith, F. P.,** Infectious feline agranulocytosis, *Am. J. Pathol.*, 16, 333, 1940.
160. **Reynolds, H. A.,** Some clinical and hematological features of virus enteritis of mink, *Can. J. Comp. Med.*, 33, 155, 1969.
161. **Jacobs, R., Weiser, M. G., Hall, R. L., and Kowalski, J. J.,** Clinicopathologic features of canine parvoviral enteritis, *J. Am. Anim. Hosp. Assoc.*, 16, 809, 1980.
162. **Johnson, R. H. and Collings, D. F.,** Experimental infection of piglets and pregnant gilts with a parvovirus, *Vet. Rec.*, 85, 446, 1969.
163. **Johnson, R. H. and Collings, D. F.,** Transplacental infection of piglets with a porcine parvovirus, *Res. Vet. Sci.*, 12, 570, 1971.
164. **Mengeling, W. L. and Cutlip, R. C.,** Reproductive disease experimentally induced by exposing pregnant gilts to porcine parvovirus, *Am. J. Vet. Res.*, 37, 1393, 1977.
165. **Joo, H. S., Donaldson-Wood, C. R., and Johnson, R. H.,** Observations on the pathogenesis of porcine parvovirus infection, *Arch. Virol.*, 51, 123, 1976.
166. **Storz, J., Leary, J. J., Carlson, J. H., and Bates, R. C.,** Parvoviruses associated with diarrhea in calves, *J. Am. Vet. Med. Assoc.*, 173, 624, 1978.

167. **Roth, S., Kaaden, O. R., van Dawen, S., and Moenning, V.,** Aleutian disease virus in B and C lymphocytes from blood and spleen and in bone marrow cells from naturally infected mink, *Intervirology*, 22, 211, 1984.
168. **Paul, P. S., Mengeling, W. L., and Brown, T. T.,** Replication of porcine parvovirus in peripheral blood lymphocytes, monocytes, and peritoneal macrophages, *Infec. Immun.*, 25, 1003, 1979.
168a. **Boosinger, T. R., Rebar, A. H., DeNicola, D. B., and Boon, G. D.,** Bone marrow alterations associated with canine parvoviral enteritis, *Vet. Pathol.*, 19, 558, 1982.
168b. **Siegl, G. and Kronauer, G.,** unpublished observations.
169. **Siegl, G. and Kronauer, G.,** A plaque assay for feline panleukopenia virus, *J. Gen. Virol.*, 46, 211, 1980.
170. **Durham, P. J. K. and Jonhson, R. H.,** Plaque titration and inhibition tests for bovine parvovirus, *J. Gen. Virol.*, 65, 1425, 1984.
171. **Cormie, J. L.,** Parvovirus vaccination, *Vet. Rec.*, 117, 418, 1985.
172. **Campbell, D. A., Jr., Manders, E. K., Oehler, J. R., Bonnard, G. D., Oldham, R. K., and Herberman, R. B.,** Inhibition of in vitro lymphoproliferative responses by in vivo passaged rat 13762 mammary adenocarcinoma cells. I. Characteristics of inhibition and evidence for an infectious agent, *Cell. Immunol.*, 33, 364, 1977.
173. **Campell, D. A., Jr., Staal, S. P., Manders, E. K., Bonnard, G. D., Oldham, R. K., Salzman, L. A., and Herberman, R. B.,** Inhibition of in vitro lymphoproliferation responses by in vivo passaged rat 13762 mammary adenocarcinoma cells. II. Evidence that Kilham rat virus is responsible for the inhibitory effect, *Cell. Immunol.*, 33, 378, 1977.
174. **Sahli, R., McMaster, G. K., and Hirt, B.,** DNA sequence comparison between two tissue-specific variants of the autonomous parvovirus, minute virus of mice, *Nucleic Acids Res.*, 13, 3617, 1985.
175. **Darrigrand, A. A., Singh, S. B., and Lang, C. M.,** Effects of Kilham rat virus on natural killer cell-mediated cytotoxicity in Brown Norway and Wistar Furth rats, *Am. J. Vet. Res.*, 45, 200, 1984.
176. **Oehler, J. R., Lindsay, L. R., Nunn, M. E., Holden, H. T., and Herberman, R. B.,** Natural cell mediated cytotoxicity in rats. II. In vivo augmentation of NK-cell activity, *Int. J. Cancer*, 21, 210, 1978.
177. **Schultz, R. D., Mendel, H., and Scott, F. W.,** Effect of feline panleukopenia virus infection on development of humoral and cellular immunity, *Cornell Vet.*, 66, 324, 1976.
178. **Carlson, J. H., Scott, F. W., and Duncan, J. R.,** Feline panleukopenia. III. Development of lesions in lymphoid tissues, *Vet. Pathol.*, 15, 383, 1978.
179. **Olsen, C. G., Stiff, M. I., and Olsen, R. G.,** Comparison of the blastogenic response of peripheral blood lymphocytes from canine parvovirus-positive and -negative outbred dogs, *Vet. Immunol. Immunpathol.*, 6, 285, 1984.
180. **Porter, D. D., Porter, H. G., and Larsen, A. E.,** Much of the increased IgG in Aleutian disease of mink is viral antibody, *J. Exp. Pathol.*, 1, 79, 1984.
181. **Perryman, L. E., Banks, K. L., and McGuire, T. C.,** Lymphocyte abnormalities in Aleutian disease virus infection of mink: decreased T lymphocyte responses and increased B lymphocyte levels in persistent viral infection, *J. Immunol.*, 115, 22, 1975.
182. **An, S. H. and Wilkie, B. N.,** Mitogen- and viral antigen-induced transformation of lymphocytes from normal mink and from mink with progressive or nonprogressive Aleutian disease, *Infec. Immun.*, 34, 111, 1981.
183. **Hahn, E. C. and Hahn, P. S.,** Autoimmunity of Aleutian disease: contribution of antiviral and anti-DNA antibody to hypergammaglobulinaemia, *Infec. Immun.*, 41, 494, 1983.
184. **Race, J. E., Bloom, M. E., and Coe, J. E.,** Demonstration of Aleutian disease virus-specific lymphocyte response in mink with progressive Aleutian disease: comparison of sapphire and pastel mink infected with different virus strains, *J. Immunol.*, 131, 1558, 1983.
185. **Burger, D., Sriranganathan, N., McDonald, T. L., and Gorham, J. R.,** Isolation of virus and antibody containing immune complexes from mink with Aleutian disease by affinity chromatography of equine complement Clq, *Am. J. Vet. Res.*, 44, 86, 1983.
186. **Porter, D. D., Dixon, F. J., and Larsen, A. E.,** Metabolism and function of gamma globulin in Aleutian disease of mink, *J. Exp. Med.*, 121, 889, 1965.
187. **Porter, D. D., Larsen, A. E., and Porter, H. G.,** The pathogenesis of Aleutian disease of mink. I. In vivo viral replication and the host antibody responses to viral antigen, *J. Exp. Med.*, 130, 575, 1969.
188. **Lodmell, D. L., Hadlow, W. J., Munoz, J. J., and Whitford, H. W.,** Hemagglutinin antibody response of normal and Aleutian disease-affected mink to Keyhole limpet hemocyanin, *J. Immunol.*, 104, 878, 1970.
189. **Kenyon, A. J.,** Immunologic deficiency in Aleutian disease of mink, *Am. J. Vet. Res.*, 27, 1780, 1966.
190. **Trautwein, G., Schneider, P., and Ernst, E.,** Untersuchungen ueber die Pathogenese der Aleutenkrankheit der Nerze. VIII. Depression der Antikoerperbildung gegen Meerrettich-Peroxidase, *Zentralbl. Veterinaermed. Reihe B*, 21, 467, 1974.
191. **Hartsough, C. F. and Gorham, J. R.,** Aleutian disease of mink, *Natl. Fur. News*, 28, 10, 1956.

192. **Lodmell, D. L. and Portis, J. L.**, Immunologic and genetic aspects of Aleutian disease, in *Immunologic Defects in Laboratory Animals*, Gershwin, M. and Merchant, B., Eds., Plenum Press, New York, 1981.
193. **Larsen, A. E. and Porter, D. D.**, Pathogenesis of Aleutian disease in mink: identification of non-persistent infections, *Infect Immun.*, 11, 92, 1975.
194. **An, S. H. and Ingram, D. G.**, Detection of inapparent Aleutian disease virus infection in mink, *Am. J. Vet. Res.*, 38, 1619, 1977.
195. **Hadlow, W. J., Race, R. E., and Kennedy, R. C.**, Comparative pathogenicity of four strains of Aleutian disease virus for pastel and sapphire mink, *Infec. Immun.*, 41, 1016, 1983.
196. **Henson, J. B., Leader, R. W., Gorham, J. R., and Padgett, G. A.**, The sequential development of lesions in spontaneous Aleutian disease of mink, *Pathol. Vet.*, 3, 289, 1966.
197. **Pan, I. C., Tsai, K. S., and Karstad, L.**, Glomerulonephritis in Aleutian disease of mink: histological and immunofluorescence studies, *J. Pathol.*, 101, 119, 1970.
198. **Drommer, W. and Trautwein, G.**, Die Pathogenese der Aleutenkrankheit der Nerze. VII. Chronische Hepatitis mit Gallengangproliferation, *Vet. Pathol.*, 12, 77, 1975.
199. **Bazeley, P. I.**, The nature of Aleutian disease in mink. I. Two forms of hypergammaglobuliemia as related to method of disease transmission and type of lesion, *J. Infect. Dis.*, 134, 252, 1976.
200. **Cheema, A., Henson, J. B., and Gorham, J. R.**, Aleutian disease of mink. Prevention of lesions by immunosuppression, *Am. J. Pathol.*, 66, 543, 1972.
201. **Porter, D. D. and Cho, H. J.**, Aleutian disease of mink: a model for persistent infection, in *Comprehensive Virology*, Fraenkel-Conrat H., and Wagner, R. R., Eds., Plenum Press, New York, 1980, 16.
202. **Porter, D. D., Porter, H. G., Suffin, S. C., and Larsen, A. E.**, Immunoglobulin classes of Aleutian disease virus antibody, *Infec. Immun.*, 43, 463, 1984.
203. **Cho, H. J. and Ingram, D. G.**, Isolation, purification and structure of Aleutian disease virus by immunological techniques, *Nature (London)*, 234, 174, 1973.
204. **Saison, R., Karstad, L., and Pridham, T. J.**, Viral plasmacytosis (Aleutian disease) in mink. VI. The development of positive coombs tests in experimental infections, *Can. J. Comp. Med.*, 30, 151, 1966.
205. **Barnett, E. V., Williams, R. C., Jr., Kenyon, A. J., and Henson, J. E.**, Nuclear antigens and antinuclear antibodies in mink sera, *Immunology*, 16, 241, 1969.
206. **Russel, A. S. and Percy, J. S.**, The antigenicity of polynucleotides in mink, rabbits, and man, *J. Rheumatol.*, 1, 66, 1974.
207. **Hahn, E. C. and Kenyon, A. J.**, Anti-deoxyribonucleic acid antibody associated with persistent infection of mink with Aleutian disease virus, *Infec. Immun.*, 29, 452, 1980.
208. **Stolze, B. and Kaaden, O.-R.**, Apparent lack of neutralizing antibodies in Aleutian disease is due to masking of antigenic sites by phospholipds, *Virology*, 158, 174, 1987.

Chapter 4

LABORATORY TESTS FOR THE DIAGNOSIS OF INFECTION WITH B19 VIRUS

B. J. Cohen

TABLE OF CONTENTS

I. INTRODUCTION

The purpose of this chapter is to outline the development and performance of diagnostic tests for B19 virus and illustrate their application with selected case histories. B19 virus was first recognized as a separate viral entity during an evaluation of laboratory tests for hepatitis B surface antigen (HBsAg).[1] A small number of sera, mainly from blood donors, were found by chance to contain a particulate antigen that was distinct from HBsAg. It reacted with certain human sera in counter-immunoelectrophoresis (CIE) and immune electron microscopy (IEM) but did not react in passive hemagglutination or radioimmunoassay (RIA) tests based on sera from sheep, horses, or guinea pigs.[1] Subsequently it transpired that B19 virus had been independently discovered in a similar way at two other centers, in France and in Japan, where donors were also tested by CIE.[2,3] In addition, other studies using CIE and IEM showed that the antibody to B19 virus was widespread in many human populations.[1,4] Initially the two techniques of CIE and IEM were the only laboratory methods available for investigating B19 virus infection, but their lack of sensitivity (CIE) or unsuitability for examining large numbers of samples (IEM) limited the scope of diagnostic studies.

The need for improved laboratory tests was met by advances on two fronts: (1) monoclonal antibodies to B19 virus were prepared permitting the development of sensitive solid phase immunoassays for viral antigen and virus-specific IgM and IgG;[5] and (2) portions of the B19 virus genome were cloned providing probes for examining clinical material for the presence of virus in DNA hybridization assays.[6,7] The latter, molecular approach has helped overcome a limitation imposed by the absence of a cell culture system for B19 virus. However, the continued failure to develop an in vitro method for propagating the virus has resulted in a chronic shortage of B19 antigen for diagnostic studies. These have therefore been confined to the relatively few centers that have obtained, frequently fortuitously, supplies of antigen from viremic blood donors.

As with other viruses the diagnosis of B19 virus infection may be made directly by detecting the virus, or indirectly by measuring a specific antibody (anti-B19) response. Since the means of isolating B19 virus in cell cultures does not yet exist, its presence must be demonstrated by other methods which often involve the assay of a viral component, viz; viral capsid antigen or nucleic acid, rather than complete virus. Specimens may be examined for antigen by one of a range of immunological methods and for nucleic acid by DNA hybridization assay. Complete virus may be demonstrated by electron microscopy but the role for this technique in routine diagnosis is limited. Measurement of the antibody response to B19 virus yields indirect evidence of infection with the virus: the presence of specific IgM class antibodies indicating recent infection and specific IgG, in the absence of anti-B19 IgM, indicating past infection and immunity to B19. The relative merits of the two diagnostic approaches, direct and indirect, will be discussed.

II. DIRECT DEMONSTRATION OF B19 VIRUS

A. Assays of B19 Virus Antigen

1. Counter-Immunoelectrophoresis

Although virtually obsolete as a diagnostic test, CIE warrants consideration because of its role in the discovery of B19 virus.[1] In the early 1970s when B19 virus was first recognized, CIE was widely used to screen blood donors for HBsAg. Of the many variations of the technique that were available, the method used by Cossart et al. in 1975[1] was based on that described by Pesendorfer and co-workers[8] in which a gel containing 0.5% agar and 0.5% agarose in a high pH, barbitone buffer is poured on to a glass slide. When it has set a double row of wells is cut in the gel. Test sera are placed in the cathodal row of wells and the opposing wells are filled with detector serum. A constant current of approximately 1 mA/

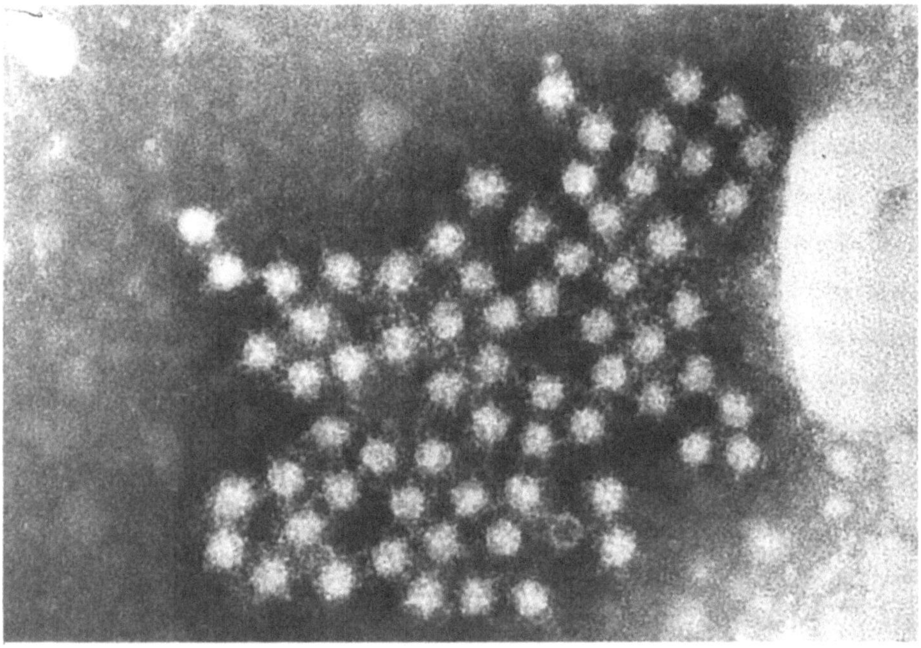

FIGURE 1. Electron micrograph of material from a precipitin line in a CIE test showing an aggregate of parvovirus particles cross-linked by antibody. The antibody has the morphology characteristic of IgG. (Magnification × 189,000.)

cm is run across the slide for 2 hr. Antigen flows with the current and is precipitated by antibody which, under the conditions of the electrophoresis, migrates from the anodal wells in the opposite, counter-current direction. A visible precipitin line is formed where the antigen and antibody reaction has taken place.

It is difficult to distinguish by eye a precipitin line formed by a B19 virus antigen/antibody reaction from that formed by HBsAg/antibody. The distinction was initially made by the different reactions of the two antigens with the human detector sera "P" and "SZ":[1] both contained antibody to HBsAg but only serum "P" contained antibody that precipitated B19 virus in CIE. It has subsequently been found by RIA tests in this laboratory that both detector sera actually contain anti-B19 IgG but the quantity in serum "P" (30 arbitrary RIA units) is greater than that in serum "SZ" (10 arbitrary RIA units).

The specificity of a CIE reaction may be confirmed by electron microscopy of the precipitin line. A small area of gel containing the line is excised with a scalpel, homogenized in a minimal quantity of distilled water, mixed with negative stain, and examined in the electron microscope. The presence of aggregates of B19 virus particles cross-linked by antibody (Figure 1) confirms that the CIE reaction is specific. The absence of such aggregates (a few scattered virus particles only may be seen) would indicate that the precipitin line was due to a nonspecific reaction.

2. Immune Electron Microscopy

The technique of IEM, pioneered in virology by Almeida,[9] enables the antigenic properties of virus particles to be studied as well as their morphology. Compared to direct electron microscopy, IEM has particular advantages when examining serum specimens for B19 virus, because parvoviruses are rather featureless small round structures which are difficult to recognize against the background of globular debris present in serum. Aggregation with

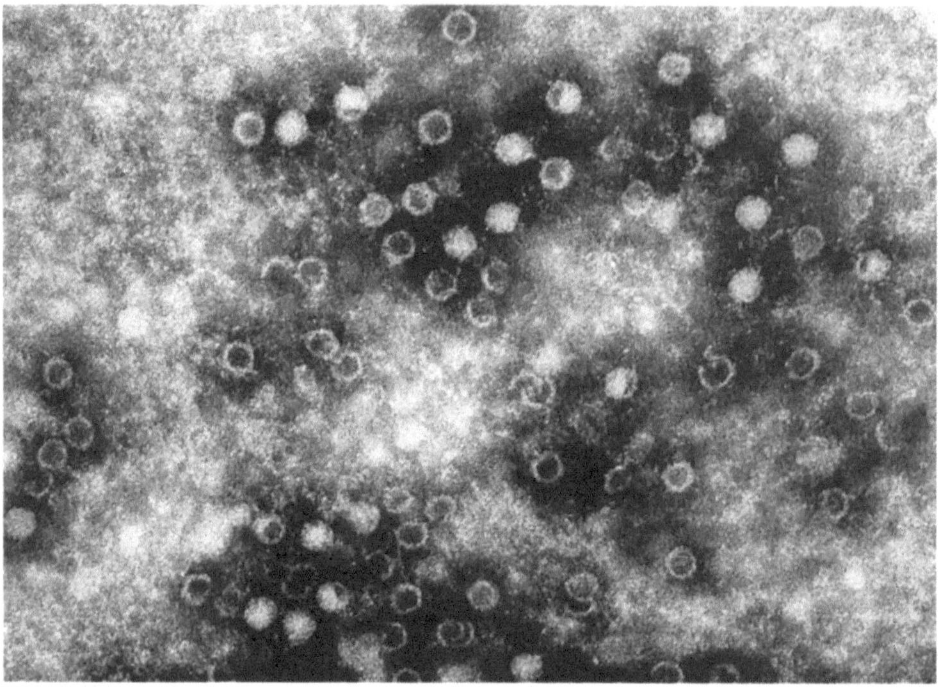

FIGURE 2. Electron micrograph of the original human parvovirus "isolate" labeled B19 showing the various morphological features characteristic of parvovirus: full, empty, part-full, and "half-shell" fragments of particles are present. (Magnification × 189,000.)

specific antibody in IEM brings into a single field of view a larger number of B19 virus particles than would be seen by direct electron microscopy. It also serves to aggregate particles with the various morphological features of a parvovirus (Figure 2), helping to establish their identity on grounds of morphology. Since the particles are aggregated by specific antibody they can also be said to have the antigenic properties of B19 virus.

In this discussion, IEM is treated as an immunological method for demonstrating B19 virus particles. Examination of virus particles without the addition of detector antibody to the test sample, i.e., direct electron microscopy, is considered below as a separate technique. As with CIE, the interest in IEM as a laboratory test for B19 virus lies largely in the part it played in the discovery of the virus. It does, however, have a continuing role in diagnosis of B19 virus infection by providing a means of confirming positive results obtained in other B19 virus assays. The visualization of distinct virus particles provides the best confirmation of the presence of virus in a sample reactive for B19 virus antigen or DNA. IEM is a labor-intensive and time-consuming technique that is suitable for examining only small numbers of specimens.

The IEM method used in this laboratory[10] involves mixing the specimen under test with an equal volume of detector anti-B19 serum. Volumes as small as 50 $\mu\ell$ usually yield a satisfactory result. The mixture is incubated for 1 hr at room temperature and then diluted in approximately 25 times the volume of phosphate buffered saline and centrifuged at 48,000 g for 1 hr. The resulting pellet is negatively stained with phosphotungstic acid and examined in the electron microscope.

When IEM was done on the sera studied by Cossart and colleagues it was found that they contained a particulate antigen that more closely resembled a parvovirus than hepatitis B virus.[1] The particles had a diameter ranging from 20.5 to 25 nm, mean 23 nm. This compares with the diameter of small round particles of HBsAg which ranges from 16 to 25 nm, mean

22 nm. Although the mean diameters of the two types of particle are very close, B19 virus particles are much more regular in size and shape than the small round particles of HBsAg which are rather pleomorphic.

There are other features of the particle morphology which distinguish B19 virus from hepatitis B virus. In addition to the small round particles of HBsAg many hepatitis B sera may contain two other morphological forms: the so-called "Dane" particle, a double-shelled structure constituting the complete virus particle,[11] and the long or filamentous forms of HBsAg. A B19 virus positive serum, on the other hand, contains only small round particles. They may be penetrated by stain, in which case they appear "empty", or they may completely exclude the negative stain, in which case they appear "full". Sometimes particles are seen that are part full and part empty. Another distinctive morphological feature commonly seen in B19 virus preparations is the "half shell" structures which are probably fragments of the viral capsid. All the morphological features of B19 are illustrated in Figure 2. HBsAg may be confused with B19 virus in IEM, however, because either the hepatitis B sera contain only small round particles (the long forms and Dane particles being absent) or the human detector serum may contain antibodies to both HBsAg and B19 virus and thus aggregate both types of particle.

3. Immunodiffusion

The application of immunodiffusion tests has been directed towards the antigenic analysis of sera known to contain B19 virus rather than the testing of unknown serum samples. It was used by Cossart and co-workers[1] to show that B19 virus gave a reaction of nonidentity with HBsAg of different subtypes. Immunodiffusion was also used by Serjeant et al.[12] to show that B19 virus found during aplastic crisis in patients with sickle cell anemia was antigenically identical to the virus detected in asymptomatic blood donors. All B19 virus "isolates" that have been tested show reactions of complete identity by immunodiffusion indicating that there is only one major antigenic type.

4. Radioimmunoassay

The RIA used to detect B19 virus antigen is a modification of the method developed to measure class-specific antibodies to B19 virus.[5] The assay is done as described below in the section on assays for antibody to B19 virus (see Section III.C), except that it is performed with a serum known to be positive for anti-B19 (either IgM or IgG). The specimen being tested for antigen is then added in place of the known B19 virus antigen reagent and the assay completed in the usual way.

RIA is a more sensitive way of detecting B19 virus antigen than CIE in two senses. It detects antigen in higher dilutions and in more specimens than CIE. In a comparison of the two methods reported previously,[5] B19 virus antigen was detected in a reference serum at a dilution of 1 in 10,000 by RIA but only at a dilution of 1 in 4 by CIE. In the same study six clinical specimens were positive for antigen by RIA but only four of them were reactive by CIE. All six were confirmed by direct or immune EM.

At present RIA probably represents the method of choice for demonstrating B19 virus antigen. It is sensitive, specific, and may be adapted to screen large numbers of specimens. However, it may be convenient for some laboratories to use CIE, since it is simpler and more rapid than RIA and the intensity of viremia in a B19 virus infection is such that a majority of virus-positive sera will prove to be reactive in the less sensitive assay. Moreover, as CIE utilizes a human convalescent serum as detector reagent it is likely to react more broadly than RIA based on monoclonal antibodies. It has been argued that all antigenic variants of B19 virus may not be detected by the RIA method.[6,7] In practice this remains a largely theoretical objection as virtually all sera in which B19 virus antigen can be detected have been reactive with monoclonal antibodies. In this laboratory, more than 50 sera with

B19 virus antigen have been examined and only one specimen failed to react with one of the commonly used monoclonal antibodies (VRL/B19/3),[13] but they all reacted with another (VRL/B19/11).[5] It would seem prudent, however, to use a combination of more than one monoclonal antibody when testing unknown specimens for B19 virus antigen by RIA.

B. Assay of B19 Virus DNA

The detection of viral nucleic acid provides an alternative approach to immunological assay of viral antigen for the demonstration of B19 virus in clinical specimens. The procedure involves immobilization of nucleic acid from the clinical material on to a nitrocellulose or nylon filter followed by hybridization with a complementary, radiolabeled DNA probe.

The two "dot-blot" hybridization assays for B19 virus DNA that have been described employ molecularly cloned portions of the viral genome as probe. In one case the cloned DNA comprised an internal Pst 1 fragment of the B19 virus genome that is approximately 700 base pairs in length.[6] In the other, approximately half the B19 virus genome was cloned by employing the technique of homopolymeric tailing.[7] Although the B19 virus clones currently described for diagnostic studies do not contain the entire genome they detect viral nucleic acid with high sensitivity suggesting that the B19 virus DNA being detected has been conserved in all virus-positive diagnostic samples found to date.[6] In other work virtually full length clones of the B19 virus genome have been obtained by separately cloning two halves of the genome which are formed on cleavage by the enzyme Bam HI,[14] but these clones have not yet been widely used in diagnostic studies.

Test samples are conveniently applied to a filter with a microfiltration apparatus. They are denatured either by mixing with an alkali solution before spotting on to the filter or by floating the filter on alkali after the specimens have been applied. The filters are then heated at 80°C for 2 hr to bind the specimens. (This step is not necessary with nylon filters). Hybridization with heat-denatured, ^{32}P-labeled probe DNA is carried out using standard procedures. After hybridization the filters are extensively washed to remove nonspecifically bound ^{32}P, dried, and then exposed to X-ray film.

A high sensitivity for "dot-blot" hybridization assay has been claimed with positive results being obtained from samples containing as little as 0.5 pg B19 virus DNA.[6] The sensitivity was found to be greater than[6] or comparable to[7] that of radioimmunoassay for viral antigen. Hybridization assays, however, have a major advantage when compared to immunological assays of B19 virus antigen. Since they are independent of antigen/antibody reactions hybridization assays can be used to detect virus in specimens where antigen reactivity is masked by the presence of antibody. This situation may occur if a sample is collected during the early recovery phase of infection when the period of viremia overlaps with the development of the specific antibody response. In such cases, CIE for viral antigen is often negative and RIA for viral antigen may also be negative (Table 1) or only weakly positive (Table 2). It is unlikely that this simply reflects a lack of sensitivity of the antigen assays because large numbers of virus particles may be seen in such specimens by direct EM (Figure 3). It is more likely that when particles are coated and aggregated by antibody they are unable to migrate through the electrophoresis gel in a CIE test, and in RIA the presence of endogenous anti-B19 attached to viral particles diminishes or eliminates antigenic sites free for binding the detector monoclonal antibody. The number of diagnostic specimens that fall into this category is difficult to estimate at present. Recently, a "dot-blot" hybridization assay has been introduced in this laboratory to examine all specimens submitted for B19 virus testing. This has produced an increased number of known virus-positive sera and studies of the diagnostic yield obtained with the hybridization assay compared to other methods for detecting B19 virus are continuing.

In addition to the detection of B19 virus in sera, DNA hybridization methods provide a sensitive means of detecting virus in other types of clinical material. Thus examination of

Table 1
FAMILY OUTBREAK OF FIFTH DISEASE

Age (years)	Sex	Illness	B19 virus antibody IgM	B19 virus antibody IgG	B19 virus Antigen (CIE and RIA)	B19 virus DNA	B19 virus EM
38	F	Rash	1.7[a]	<1	Negative	Positive	Positive
9	F	Rash	15	45	Negative	Negative	N.T.
4	F	None	19	7.2	Negative	Negative	N.T.
1 1/2	F	None	1.3	<1	Negative	Positive	Positive

Note: N.T. = not tested.

[a] Arbitrary RIA units. Specimens provided by Dr. M. H. Hambling, Public Health Laboratory, Leeds, England.

Table 2
A FAMILY OUTBREAK OF APLASTIC CRISIS

Age (years)	Sex	Illness		B19 virus antibody IgM	B19 virus antibody IgG	B19 virus Antigen (CIE)	B19 virus Antigen (RIA)	B19 virus DNA	B19 virus EM
12	M	HS/Aplastic crisis	a	1.2[a]	<1	–	+	Positive	Positive
			b	>100	27.4	–	–	N.T.	N.T.
17	F	HS/aplastic crisis	a	1.9	<1	–	–	Positive	Positive
			b	>100	31.0	–	–	N.T.	N.T.
14	M	None	a	7.4	1.6	–	–	Positive	Positive
			b	35.8	28.0	–	–	N.T.	N.T.
16	M	HS/aplastic crisis	a	>100	53.0	–	–	Positive	Positive
			b	29.0	76.0	–	–	N.T.	N.T.
12	F	HS/aplastic crisis	a	14.0	27.0	–	–	Positive	Positive
74	F	None	a	<1	17.5	–	–	Positive	Positive
			b	<1	22.0	–	–	N.T.	N.T.
36	M	None	a	<1	5.3	–	–	Positive	Positive
			b	<1	7.4	–	–	N.T.	N.T.

Note: HS = Hereditary spherocytosis; a = acute serum 23/4/79, b = convalescent serum 4/5/79; N.T. = not tested.

[a] Arbitrary RIA units. Specimens provided by Dr. T. C. Harvey, Wallsall Manor Hospital, England.

throat swab specimens revealed the presence of virus in the respiratory tract, indicating that the infection was transmitted by this route.[15] This was confirmed during experiments on human volunteers[16] when it was also shown by hybridization assay that B19 virus was not excreted in feces or in urine.

DNA hybridization assays now represent the optimal method for demonstrating B19 virus in clinical specimens. It overcomes problems inherent in immunological assays of viral antigen and can be readily applied to screen large numbers of specimens.[6,7] The disadvantage of the present hybridization methods for B19 virus DNA detection is that they employ radioactive probes. It will be more satisfactory when alternative methods can be introduced using nonradioactive probes based on the biotin-avidin system[17] for example.

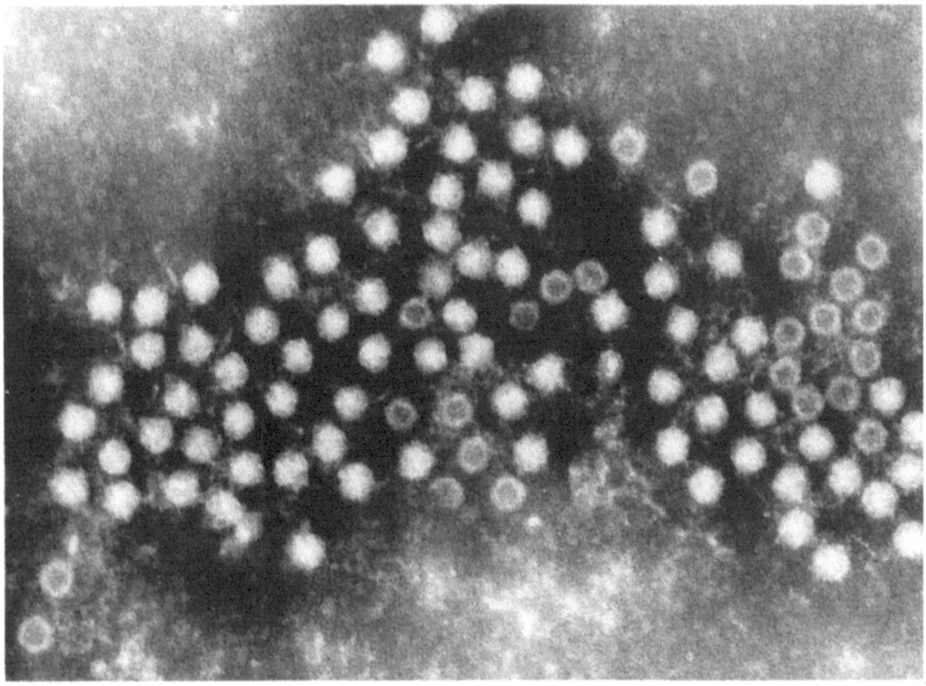

FIGURE 3. Direct electron microscopy of serum containing B19 virus particles cross-linked by antibody of the IgM class; no exogenous antibody was added. (Magnification × 189,000; courtesy Miss P. A. Litton.)

C. Direct Electron Microscopy

The intensity of viremia in a B19 virus infection is such that electron microscopic techniques, both immune and direct EM, provide an effective means of detecting the virus. It can be estimated that a virus-positive serum may contain in excess of 10^{10} particles per milliliter. This estimate is derived from the B19 virus DNA content of virus-positive serum, which may be measured in microgram quantities,[18] and the figure quoted by Anderson et al.[6] that 0.5 pg of B19 virus DNA is equivalent to 10^4 parvovirus particles.

Direct EM is especially valuable as a confirmatory procedure to examine specimens thought to contain B19 virus because they are positive in hybridization assays for the viral DNA. It is not always possible to detect viral antigen in B19 virus-positive specimens because specific antibody may also be present, masking the antigen and rendering it undetectable by RIA or CIE (Table 1). If such specimens are centrifuged under the conditions described in the section dealing with IEM, but without the addition of exogenous anti-B19, and the resulting pellets examined by electron microscopy, they may be seen to contain aggregates of parvovirus particles cross-linked by antibody (Figure 3). Very often the cross-linking antibody has the morphology characteristic of IgM (Figure 3). Virus-antibody complexes are found when viremia clears and B19 virus antibody develops[16] and may have an etiological role in the pathogenesis of the rash and joint pains that occur at this phase of infection.

An alternative method for examining clinical material for parvovirus particles was employed by Appleton and co-workers during a study of fecal specimens from patients with gastroenteritis.[19] They fractionated the specimens on CsCl density gradients and examined the fractions by direct EM. The density gradient serves both to purify and concentrate virus particles, separating them from other material in the fecal specimen which might otherwise obscure them in electron microscopy. It has not generally been necessary to adopt this approach for B19 virus because the intensity of the viremia is such that a concentration step to help visualize the virus particles in serum is not necessary.

III. DEMONSTRATION OF SPECIFIC ANTIBODIES TO B19 VIRUS

A. Counter-Immunoelectrophoresis

Initially CIE was the most widely applied technique for detection of antibody to B19 virus (anti-B19). It was used to show that anti-B19 was not only common in human populations,[1] but was principally acquired in childhood.[20] CIE tests for anti-B19 also played an important part in establishing the link between B19 virus infection and the aplastic crisis of sickle cell anemia, being used to demonstrate seroconversion in patients in whom the virus had not been found.[12,21] Standard CIE methods may be used to detect anti-B19 as outlined for the assay of B19 virus antigen except that test samples are electrophoresed against serum containing antigen. This reagent is generally used without dilution and CIE therefore consumes relatively large amounts of the antigen reagent which is in such limited supply.

Early antibodies to B19 virus are efficiently detected by CIE. In one study, all acute phase sera from 33 children with typical erythema infectiosum were reactive for anti-B19 by CIE.[22] Anti-B19 IgM undoubtedly makes a large contribution to these reactions since this class of antibody readily precipitates in CIE tests.[23] In the absence of anti-B19 IgM a relatively high level of anti-B19 IgG is required to react in CIE: many sera with <10 RIA units of anti-B19 IgG are negative by CIE.[5] As the anti-B19 response wanes, therefore, it may become undetectable by CIE and this may occur 12 to 24 months after infection.[21] This phenomenon probably accounts for the difference in the anti-B19 prevalence in blood donors when measured by CIE (43%) and RIA (61%).[5]

In spite of some advantages CIE has been largely superceded by RIA for the detection of antibodies to B19 virus. Though it is a simple and rapid test CIE is uneconomic in the use of antigen. While it is feasible to use CIE for diagnostic work as it detects early antibodies well, it does not discriminate IgM from IgG and a firm diagnosis can only be made by demonstrating a seroconversion with paried sera. Because it lacks the sensitivity to detect small quantities of anti-B19 IgG CIE is inadequate for the assessment of immunity to B19 virus and for sero-epidemiological studies.

B. Immune Electron Microscopy

IEM has been applied to detect anti-B19 in only one study where it was used to demonstrate that B19 virus was serologically distinct from fecal parvo-like virus.[24] Like CIE, IEM generally detects total antibody and is rather uneconomic in the use of antigen. However, IEM is said to be more sensitive than CIE and can be used to demonstrate IgM and IgG class anti-B19 in serum fractions.[24]

C. Radioimmunoassay

Unlike the antibody assays described above, RIA techniques have been principally used to measure IgM and IgG class antibodies to B19 virus. This has great value for diagnostic studies because it permits diagnosis of recent infection by demonstration of anti-B19 IgM in a single acute phase serum. Tests for anti-B19 IgM were first introduced by Anderson et al.,[25] who developed an "IgM capture" RIA (MACRIA) that was based on an antibody reagent consisting of the IgG fraction of an immune human serum. This failed to discriminate as clearly between positive and negative specimens as MACRIA systems for other viruses based on either human[26,27] or hyperimmune animal[28,29] sera. A shortage of antigenic material precluded the preparation of a hyperimmune animal serum to B19 virus but a suitable antibody reagent was provided by application of the hybridoma technique of Kohler and Milstein.[30] This was used to prepare monoclonal antibodies to B19 virus which permitted the development of sensitive and specific tests that fully exploited the advantages of the MACRIA system.[5] In addition to an assay of anti-B19 IgM, an analogous test for anti-B19 IgG was developed based on "IgG capture".

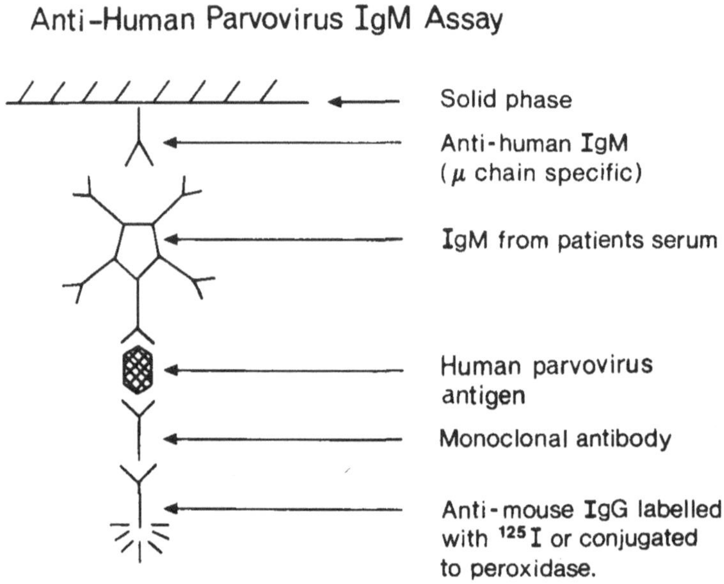

FIGURE 4. Schematic representation of the assay for anti-B19 IgM. The design of the assay can be altered to detect anti-B19 IgG or B19 virus antigen.

The general features of antibody capture assays have been reviewed by Schmidt,[31] and their development for diagnosis of B19 virus infection described by Cohen et al.[5] In the first stage of the MACRIA test, IgM from the patient's serum is "captured" on to the solid phase by anti-human IgM (Figure 4). (In the analagous capture assay for anti-B19 IgG, this class of antibody is captured by anti-human IgG). To determine whether any of the captured antibody has the sought-after specificity, B19 virus antigen and monoclonal anti-B19 are applied. These two reagents maybe added sequentially or simultaneously. The final stage of the assay is accomplished by adding a labeled anti-mouse IgG reagent. Attempts in this laboratory to develop the test by adding labeled monoclonal anti-B19 have not been successful. Although the system can be set up as an enzyme linked immunosorbent assay (ELISA) using anti-mouse IgG conjugated to peroxidase, RIA has generally been preferred using anti-mouse IgG labeled with ^{125}I as this gives satisfactory results with a smaller quantity of B19 antigen than that required for ELISA.

In MACRIA tests, the concentration of specific IgM is quantified in terms of arbitrary units as described by Tedder and Wilson-Croome for hepatitis B core IgM.[27] This gives a value for the proportion of the IgM fraction that is specific for the antigen in the assay. In the MACRIA for anti-B19 IgM a value of >1 unit is usually considered significant, i.e., diagnostic of recent infection.[5] Low values, however, need to be interpreted with care as nonspecific factors, especially rheumatoid factor, may occasionally give rise to weak false positive results.[5] It has also been reported that high levels of rubella IgM may give rise to trace reactions for anti-B19 IgM in MACRIA tests, and vice versa, adding to the diagnostic confusion between the two infections.[32] The interpretation of weakly positive anti-B19 IgM results may be aided by the result of assay for anti-B19 IgG, which is usually in the range of 50 to 100 units following recent infection; or by testing a second, convalescent serum though this vitiates the diagnostic approach of anti-B19 IgM testing on a single, acute serum.

Whereas MACRIA tests are frequently employed in diagnostic virology, antibody capture has not been widely used for measurement of specific IgG. The technique is less sensitive for virus-specific IgG than IgM because the proportion of virus-specific antibody in the IgG fraction is smaller than that in the IgM fraction.[31] The application of antibody capture to

B19 virus testing represents the first major use of the technique for assay of virus-specific IgG and has arisen because there is a lack of satisfactory alternatives. The disadvantages of CIE and IEM have already been discussed. Neutralization tests for anti-B19 are not possible in the absence of a cell culture system and, unlike most other parvoviruses, B19 virus does not agglutinate red blood cells[1] which rules out hemagglutination inhibition assays. Other RIA procedures for anti-B19 have been examined in this laboratory but were not successful. Competitive RIA, which detects total antibody, has not been developed for B19 virus testing because neither monoclonal anti-B19 nor convalescent human serum provides a suitable antibody reagent for this assay and hyperimmune animal sera are not available. Indirect RIA for anti-B19 IgG is problematic because the antigen in the test is itself derived from human serum and contains traces of IgG that react with the labeled antiglobulin serum used in this kind of assay. Antibody capture therefore provides the best approach to anti-B19 IgG testing at present but its efficacy is difficult to assess because alternative methods of comparable sensitivity are not available.

IV. SELECTED CASE HISTORIES

A. Family Outbreak of Fifth Disease

Four members of a family were investigated during a community-wide outbreak of *fifth disease* that occurred in a small rural town in the north of England during April and May 1985. The mother and one child had an illness with a rash but two younger children were asymptomatic (Table 1). Anti-B19 IgM was detected at diagnostic levels in the child with a rash and in one of the asymptomatic children, and their sera also contained anti-B19 IgG. The mother and the other asymptomatic child had only trace amounts of anti-B19 IgM and were anti-B19 IgG negative. However, their sera were positive for B19 virus DNA, though tests for B19 virus antigen were negative. Direct EM of these specimens revealed the presence of aggregates of parvovirus particles cross-linked by antibody similar to those shown in Figure 3.

The pattern of laboratory results in this family outbreak shows that recent infection with B19 virus can be proved serologically in patients with a rash and in those without symptoms by demonstrating anti-B19 IgM supported by the finding of anti-B19 IgG. Equally, B19 virus infection can be shown in both symptomatic and asymptomatic cases by detection of viral DNA. Although the presence of virus could not be proved by tests for viral antigen it was confirmed by direct EM. The coating of virus particles by endogenous antibody may have rendered them nonreactive in the antigen assays.

B. Family Outbreak of Aplastic Crises

An outbreak of *aplastic crises* in a family affected by hereditary spherocytosis (HS) was observed in a town in the midlands of England in April 1979.[5] B19 virus infection was diagnosed serologically on paired sera from four siblings with HS and from one sibling unaffected by HS who remained well as did the two adults in the family who were tested. The sera were stored and reexamined when a DNA hybridization assay became available.[7]

The serological findings on this family (Table 2) illustrate diagnosis of B19 virus infection in the children by demonstration of the virus (Case 1), by an anti-B19 IgM response (Cases 1 to 5), and by seroconversion (Cases 1 and 2). In the two adults (Cases 6 and 7) stable low levels of anti-B19 IgG were found in the paired sera indicating immunity from a past B19 virus infection. The presence of virus in acute phase serum from Case 1 was confirmed by direct EM (Figure 3) and by DNA hybridization assay.

C. Probable Viral Encephalitis in Two Siblings

An 8-year-old boy with a history of drowsiness, vomiting blood, pyrexia, and sweating

Table 3
PROBABLE VIRAL ENCEPHALITIS IN TWO CHILDREN

Age (years)	Sex	Specimen	B19 virus antibody IgM	B19 virus antibody IgG	B19 virus Antigen (CIE)	B19 virus Antigen (RIA)	B19 virus DNA	B19 virus EM
8	M	Serum 5/7/85	<1[a]	<1	–	–	–	N.T.
		Serum 15/7/85	<1	<1	+	+	+	+
		Serum 16/7/85	1.2	<1	–	+	+	+
(Onset of illness) 4/7/85		Serum 18/9/85	6.8	>100	–	–	–	N.T.
		Feces 15/7/85	N.T.	N.T.	–	–	–	N.T.
		Throat 5/7/85 Swab	N.T.	N.T.	–	–	–	N.T.
5	M	Serum 15/7/85	<1	<1	–	–	–	N.T.
		Serum 4/9/85	8.0	6.0	–	–	–	N.T.
(Onset of illness) 15/7/85		Rectal 15/7/85 Swab	N.T.	N.T.	–	–	–	N.T.
		Throat & nose 15/7/85 Swab	N.T.	N.T.	–	–	–	N.T.

Note: N.T. = not tested.

[a] Arbitrary RIA units. Specimens provided by Dr. E. A. Richardson, Public Health Laboratory, Poole, England.

was admitted to hospital in a town on the south coast of England in July 1985 during a period when B19 virus infection was widespread in that area.[33] His condition deteriorated; he became semicomatose and was transferred to the intensive care unit. He began to improve after a week but his knees and ankles were noticed to be puffy and warm and his face pink. After a further 10 days he was well enough to be discharged home. A 5-year-old male sibling was admitted to hospital with a similar though less severe illness 11 days after his brother. Diagnoses of probable viral encephalitis were made.

On admission to the hospital neither boy had any evidence of B19 virus infection, but during convalescence the virus was found in the elder boy's serum. At first it could be detected by both CIE and RIA for B19 virus antigen (Table 3). B19 virus DNA was also found and direct EM showed there were scattered parvovirus particles that were not coated by antibody (Figure 5). Addition of specific antibody in immune EM produced aggregates of particles as shown in Figure 1. On the next day antigen was undetectable by CIE though RIA remained positive and a trace of anti-B19 IgM was present. The assay for B19 virus DNA was still strongly positive but direct EM showed that some of the parvovirus particles were now coated and aggregated by antibody as in Figure 3. Acute phase serum from the younger boy was not examined but convalescent sera from both boys showed low levels of anti-B19 IgM and high levels of anti-B19 IgG.

Other likely causes of viral encephalitis were excluded in these cases,[33] but in spite of the ample evidence of B19 virus infection it is difficult to ascribe causal significance to its presence. There was no cerebrospinal fluid available for B19 virus tests to assess the involvement of the virus in the neurological features of the illness. Virus was not recovered from throat swab or fecal specimens, but was found in sera taken after the onset of symptoms. It is probable, therefore, that evidence of B19 virus infection was a chance finding, occurring during a period when the virus was known to be epidemic.

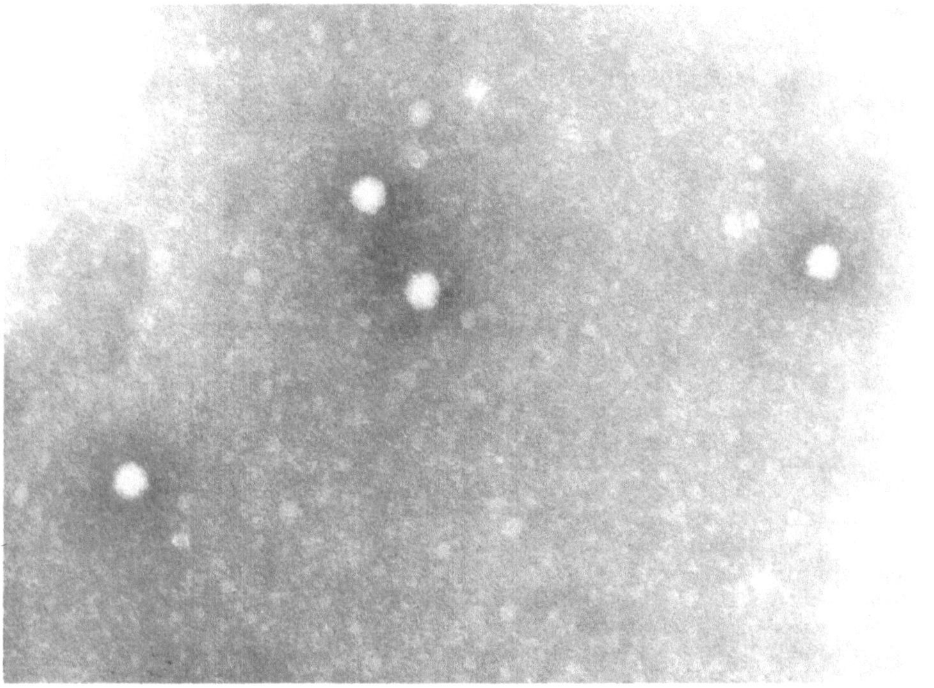

FIGURE 5. Electron micrograph of scattered parvovirus particles not coated by antibody. (Magnification × 189,000; courtesy Miss P. A. Litton.)

V. CONCLUSIONS

Infection with B19 virus was recognized by relatively simple and insensitive laboratory tests which have now been replaced by more elaborate and sensitive methods that exemplify the attention paid to rapid virus diagnosis in recent years. Detection of the virus, which initially relied on immunological techniques or electron microscopy, has now been advanced by the application of DNA hybridization assays. Similarly, detection of antibodies to B19 virus has been enhanced by the application of antibody capture techniques with monoclonal antibodies. The new laboratory methods have helped achieve a greater understanding of the pathogenic effects of B19 virus.

In spite of improved methods, diagnosis by direct demonstration of B19 virus is intrinsically limited because the viremic phase of infection is brief and often precedes the onset of symptoms. On the other hand, demonstration of specific antibodies, especially of the IgM class, is rewarding for diagnostic studies, because anti-B19 IgM is detectable for 2 to 3 months and is usually present when the illness develops. Examination of clinical specimens for virus, however, remains valuable especially in patients with chronic hemolytic disorders who may present in aplastic crisis at an early stage of infection. The collection of virus-positive material from such cases and from other clinical manifestations of infection will enable comparison to be made of B19 virus "strains" by DNA analysis with restriction enzymes or by antigenic analysis with monoclonal antibodies.

Future developments of laboratory tests for B19 virus are likely to include the provision of nonradioactive methods for both DNA hybridization and antibody capture assays. This should be feasible for the DNA assay where an unlimited supply of diagnostic reagent exists in the form of molecularly cloned B19 virus DNA sequences. Future development of antibody assays is, however, constrained by the lack of antigenic material. This may be overcome by finding a cell culture system in which B19 virus can be propagated or by expression of the genes coding for viral antigen through application of recombinant DNA techniques. The large demand for diagnostic B19 virus tests which has built-up in recent years certainly requires that this problem be solved and that laboratory tests for B19 virus be made more widely available.

REFERENCES

1. **Cossart, Y. E., Field, A. M., Cant, B., and Widdows, D.,** Parvovirus-like particles in human sera, *Lancet,* 1, 72, 1975.
2. **Courouce, A. M., Ferchal, F., Muller, A., Soulier, J. P., Morinet, F., Drouet, J., and Perol, Y.,** Human parvovirus infections in France, *Lancet,* 1, 160, 1984.
3. **Okochi, K., Mori, R., Miyazaki, M., Cohen, B. J., and Mortimer, P. P.,** Nakatani antigen and human parvovirus (B19), *Lancet,* 1, 160, 1984.
4. **Paver, W. K. and Clarke, S. K. R.,** Comparison of human faecal and serum parvo-like viruses, *J. Clin. Microbiol.,* 4, 67, 1976.
5. **Cohen, B. J., Mortimer, P. P., and Pereira, M. S.,** Diagnostic assays with monoclonal antibodies for the serum parvovirus-like virus (SPLV), *J. Hyg.,* 91, 113, 1983.
6. **Anderson, M. J., Jones, S. E., and Minson, A. C.,** Diagnosis of human parvovirus infection by dot-blot hybridisation using cloned viral DNA, *J. Med. Virol.,* 15, 163, 1985.
7. **Clewley, J. P.,** Detection of human parvovirus using a molecularly cloned probe, *J. Med. Virol.,* 15, 173, 1985.
8. **Pesendorfer, F., Krassnitsky, O., and Wewalka, F.,** Immunoelectrophoretic methods, *Bull. W. H. O.,* 42, 974, 1970.
9. **Almeida, J. D. and Waterson, A. P.,** The morphology of virus-antibody interaction, *Adv. Virus Res.,* 15, 307, 1969.
10. **Field, A. M.,** Diagnostic virology using electron microscopic techniques, *Adv. Virus Res.,* 27, 1, 1982.
11. **Dane, D. S., Cameron, C. H., and Briggs, M.,** Virus-like particles in serum of patients with Australia-antigen-associated hepatitis, *Lancet,* 1, 695, 1970.
12. **Serjeant, G. R., Topley, J. M., Mason, K., Serjeant, B. E., Pattison, J. R., Jones, S. E., and Mohamed, R.,** Outbreak of aplastic crisis in sickle cell anaemia associated with parvovirus-like agent, *Lancet,* 2, 595, 1981.
13. **Caul, E. O. and Cohen, B. J.,** unpublished observations.
14. **Cotmore, S. F. and Tattersall, P.,** Characterisation and molecular cloning of a human parvovirus genome, *Science,* 226, 1161, 1984.
15. **Plummer, F. A., Hammond, G. W., Forward, K., Selka, L., Thompson, L. M., Jones, S. E., Kidd, I. M., and Anderson, M. J.,** An erythema infectiosum-like illness caused by human parvovirus infection, *N. Engl. J. Med.,* 313, 74, 1985.
16. **Anderson, M. J., Higgins, P. G., Davis, L. R., Willman, J. S., Jones, S. E., Kidd, I. M., Pattison, J. R., and Tyrrell, D. J.,** Experimental parvoviral infections in humans, *J. Infect. Dis.,* 152, 257, 1985.
17. **Leary, J. L., Brigati, D. J., and Ward, D. C.,** Rapid and sensitive colorimetric method for visualising biotin-labeled DNA probes hybridised to DNA or RNA immobilised on nitrocellulose: bio-blots, *Proc. Natl. Acad. Sci. U.S.A.,* 80, 4045, 1983.
18. **Summers, J., Jones, S. E., and Anderson, M. J.,** Characterisation of the genome of the agent of erythrocyte aplasia permits its classification as a human parvovirus, *J. Gen. Virol.,* 64, 2567, 1983.
19. **Appleton, H., Buckley, M., Thom, B. T., Cotton, J. L., and Henderson, S.,** Virus-like particles in winter vomiting disease, *Lancet,* 1, 409, 1977.
20. **Edwards, J. M. B., Kessel, I., Gardner, S. D., Eaton, B. R., Pollock, T. M., Fleck, D. G., Gibson, P., Woodroof, M., and Porter, A. D.,** A search for a characteristic illness in children with serological evidence of viral or toxoplasma infection, *J. Infect.,* 3, 316, 1981.
21. **Anderson, M. J., Davis, L. R., Hodgson, J., Jones, S. E., Murtaza, L., Pattison, J. R., Stroud, C. E., and White, J. M.,** Occurrence of infection with a parvovirus-like agent in children with sickle cell anaemia during a two-year period, *J. Clin. Pathol.,* 35, 744, 1982.
22. **Nunoue, T., Okochi, K., Mortimer, P. P., and Cohen, B. J.,** Human parvovirus (B19) and erythema infectiosum, *J. Pediatr.,* 107, 38, 1985.
23. **Cohen, B. J.,** The IgM antibody responses to the core antigen of hepatitis B virus, *J. Med. Virol.,* 3, 141, 1978.

24. **Paver, W. K. and Clarke, S. K. R.,** Comparison of human and serum parvo-like viruses, *J. Clin. Microbiol.*, 4, 67, 1976.
25. **Anderson, M. J., Davis, L. R., Jones, S. E., and Pattison, J. R.,** The development and use of an antibody capture radioimmunoassay for specific IgM to a human parvovirus-like agent, *J. Hyg.*, 88, 335, 1982.
26. **Flehmig, B., Rauke, M., Berthold, H., and Gerth, H. J.,** A solid-phase radioimmunoassay for detection of IgM antibodies to hepatitis A virus, *J. Infect. Dis.*, 140, 169, 1979.
27. **Tedder, R. S. and Wilson-Croome, R.,** Detection by radioimmunoassay of IgM class antibody to hepatitis B core antigen: a comparison of two methods, *J. Med. Virol.*, 6, 235, 1980.
28. **Tedder, R. S., Yao, J. L., and Anderson, M. J.,** The production of antibodies to rubella haemagglutinin and their use in antibody capture assays for rubella-specific IgM, *Journal of Hyg.*, 88, 335, 1982.
29. **Brown, D. W. G., Gardner, S. D., Gibson, P. E., Field, A. M.,** BK virus specific IgM responses in cord sera, young children and healthy adults by RIA, *Arch. Virol.*, 82, 149, 1984.
30. **Kohler, G. and Milstein, C.,** Derivation of antibody-producing tissue culture and tumour lines by cell fusion, *Eur. J. Immunol.*, 6, 577, 1976.
31. **Schmidt, N.,** Update on class-specific viral antibody assays, *Clin. Immunol. Newsl.*, 5, 81, 1984.
32. **Kurtz, J. B. and Anderson, M. J.,** Cross-reactions in rubella and parvovirus specific IgM tests, *Lancet*, 2, 1356, 1985.
33. **Richardson, E. A. and Cohen, B. J.,** unpublished observations.

Chapter 5

B19 VIRUS INFECTION AND THE APLASTIC CRISIS

G. R. Serjeant and A. R. Goldstein

TABLE OF CONTENTS

The aplastic crisis is an acute self-limited erythropoietic arrest. It is a feature common to chronic hemolytic disorders, and has been reported in hereditary spherocytosis, acquired autoimmune hemolytic anemia, paroxysmal nocturnal hemoglobinuria, enzymopathies associated with shortened red cell survival, and hemoglobinopathies. Sickle cell disease, largely because of the frequency of this condition, has contributed most to the understanding of the epidemiology of the aplastic crisis.

I. HISTORICAL OBSERVATIONS

The aplastic crisis in sickle cell disease was first recognized by Singer et al.[1] who reported an acute fall in hemoglobin from 8.6 to 3.5 g/dℓ during a vague respiratory illness in a 9-year-old boy. The reticulocyte count was 0.2% and marrow examination revealed extreme depression of red cell precursors. Shortly afterward, normoblasts and then reticulocytes increased dramatically in the peripheral blood, the bone marrow showed intense erythropoietic hyperplasia, and the hemoglobin rose to its original level over 3 weeks. His 11-year-old sister with sickle cell disease was admitted to the hospital with similar symptoms on the same day. This report contained many of the cardinal features of the aplastic crisis.

There followed a series of reports describing an essentially similar picture.[2-7] Common features included an association with a vague upper respiratory tract illness, its epidemic nature, and the tendency to familial cases. At that time the etiology was unknown and MacIver and Parker-Williams[6] searched for a variety of bacterial and viral agents but were unable to identify a common pathogen. Following a chance discovery that an aplastic crisis in a child with homozygous sickle cell (SS) disease in London was associated with seroconversion to B19 virus, Pattison et al.[8] reported that six cases of aplastic crisis in SS disease had evidence of recent B19 virus infection. This association was further confirmed by analysis of an epidemic of aplastic crises in Jamaica between 1979 to 1981, in which 24 (86%) of 28 cases occurred concurrently with B19 virus infection.[9]

This strong association of a specific parvovirus infection and the aplastic crisis is not confined to SS disease but has been reported in hereditary spherocytosis,[10] beta thalassemia intermedia,[11] and pyruvate kinase deficiency.[12]

II. DEFINITION OF THE APLASTIC CRISIS

The hematological features of SS disease are maintained at steady-state levels by moderate increases in erythropoietic volume and activity. A variety of infective, metabolic, and nutritional disorders may impair this erythropoietic response leading to a relative marrow hypoplasia and hemoglobin and reticulocyte values falling below steady-state levels. These hypoplastic events are common in pneumonia and septicemia, they rarely if ever reflect the effects of B19 virus infection, and must be clearly differentiated from the aplastic crisis.

The aplastic crisis is a self-limited cessation of predominantly erythropoietic activity associated with a dramatic fall in hemoglobin level and an absence of reticulocytes from the peripheral blood. The presence of more than 1% reticulocytes makes the diagnosis of aplastic crisis unlikely, unless there is a daily dramatic increase in reticulocyte count consistent with the recovery phase of the aplastic crisis. When defined in this way, there is a close association of these events with concurrent B19 virus infection.

III. HEMATOLOGICAL FEATURES OF THE APLASTIC CRISIS

The principle hematological features of a classic aplastic crisis in SS disease are shown in Figure 1. Following cessation of erythropoietic activity (day 0), the hemoglobin level falls steadily at a rate determined by the red cell half life and to a level determined by this

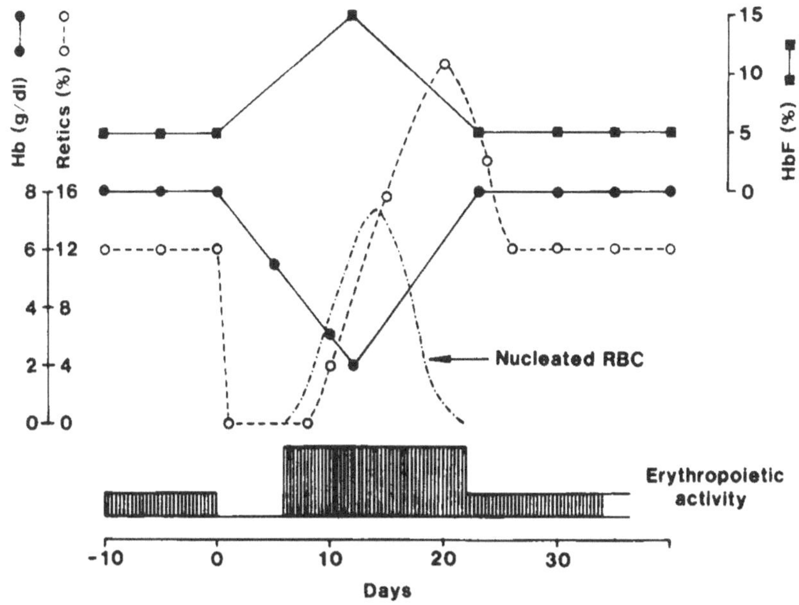

FIGURE 1. Schematic representation of some of the hematological events during an aplastic crisis.

rate and by the duration of aplasia. Because the period of erythropoietic aplasia is generally 7 to 10 days, patients with relatively long surviving red cells may show only modest falls in hemoglobin. Thus patients with sickle cell-hemoglobin C (SC) disease, sickle cell-beta$^+$ thalassemia, and some patients with SS disease or sickle cell-beta^0 thalassemia with long red cell survival may show hemoglobin falls of only 2 to 4 g/dℓ and are therefore unlikely to present with clinical features of acute anemia. On the other hand, in patients with SS disease or sickle cell-beta^0 thalassemia with short red cell survival, the hemoglobin may fall quickly to life-threatening levels of 1 to 2 g/dℓ.

The Jamaican cohort study, a representative group of patients with sickle cell disease ascertained by neonatal screening and followed prospectively, has been a valuable source of observations on the aplastic crisis. In 62 cases of B19 virus-related aplasias in these children, the average fall in hemoglobin level was 3.9 g/dℓ (range 1.7 to 6.9 g/dℓ) and the mean level at clinical presentation was 3.9 g/dℓ (range 1.8 to 8.5 g/dℓ).

Reticulocytes which disappear from the peripheral blood at the onset of aplasia reappear in large numbers when erythropoietic activity is resumed after 7 to 10 days and are generally preceded by an outpouring of nucleated red blood cells. In the cohort study, reticulocytes were absent from the peripheral blood in 43 cases (69%), and were between 0.1 to 1.4% in 14 cases (23%) at presentation. In five patients, reticulocyte counts were higher (6 to 25%) when first seen and all manifested a subsequent pattern compatible with the recovery phase.

Fetal hemoglobin (HbF) levels also change during the aplastic crisis in SS disease. Levels of HbF are generally elevated in steady-state SS disease but the amount of HbF varies markedly between red cells. Erythrocytes containing large amounts of HbF and hence smaller amounts of HbS, are less likely to undergo intravascular sickling and therefore survive longer in the circulation. This prolonged survival of high HbF-containing cells is reflected in a gradual increase in total HbF during the aplastic crisis which falls as soon as erythropoietic activity is resumed and low HbF containing erythrocytes are released again.[6] In the cohort study, HbF levels at the time of presentation with aplasia were elevated in 22/25 (88%) in those of whom these were measured, the average increase being 58% (range 10 to 133%).

During aplasia, the primary marrow failure is reflected in high erythropoietin levels[14-16] and the lack of utilization of iron results in high serum iron levels.

The effects of aplasia are not limited to the erythropoietic series and a depression of platelet count has been previously reported.[17] Among cohort study patients, the platelet count fell in 27/37 (73%) in whom measurements were available, by an average of 201,000 (range 62,000 to 584,000) compared to the child's own steady-state values. Since steady-state counts are higher than in normal people this fall lacks clinical significance and only four children had levels below 150,000.

The intense erythropoietic activity occurring during recovery from aplasia causes a rapid consumption of hematinics and the magnitude of the marrow response may be limited by the availability of iron and/or folic acid. In a typical case the hemoglobin climbs to its previous steady-state value over 2 to 3 weeks.

It is important to realize that the aplastic crisis is not confined to patients with chronic hemolytic disease but also occurs in hematologically normal persons (see Chapter 7). However, with the markedly shortened red cell survival characterizing hemolytic disease, events in the bone marrow are rapidly reflected in the peripheral blood, whereas in hematologically normal people there may be little change. Thus, in experimental B19 virus infection of susceptible volunteers,[18] reticulocytes disappeared from the peripheral blood 8 to 10 days after infection and remained absent for 7 to 10 days although the hemoglobin levels only fell by a daily average of 1.4%.

IV. CLINICAL FEATURES OF THE APLASTIC CRISIS

The clinical features are dominated by pallor, weakness, and lethargy secondary to the falling hemoglobin levels but there are in addition, a variety of nonspecific manifestations. In the cohort study cases, the majority had a prodomal illness which included fever, abdominal pain, headache, upper respiratory tract infection, anorexia, and vomiting.

On examination, pallor was the most constant finding with signs of cardiac failure such as tachycardia, hepatomegaly, and gallop rhythm being not uncommon. A pyrexia occurred in 68% and exceeded 38.0°C in 23 (37%) of the children. However, despite these features a clinical diagnosis of aplastic crisis was only suspected in half the children prior to hematological investigation.

B19 virus infection is the cause of erythema infectiosum (Chapter 6) and volunteer studies[18] identified two forms of skin involvement following experimental infection. Coincident with the viremia 1 week after inoculation there were nonspecific symptoms including itching, and 17 to 18 days after inoculation there was a fine maculopapular rash typical of erythema infectiosum in adults.

It was therefore of interest that despite close questioning and observation, skin manifestations appeared rare in the aplastic crisis. Among the 62 Jamaican cases, generalized pruritis without rash occurred at the time of presentation with aplasia in one child aged 6 years and a macular rash occurred on the face and trunk 1 week prior to aplasia in a $7^1/_2$-year-old child. The former event is typical of that seen in volunteers at the time of viremia but the latter is not typical of any recognized consequences of B19 virus infections.

The reasons for the infrequency of skin manifestations are unknown. It is highly unlikely that the dark skin causes these changes to be overlooked and it is possible that racial differences affect expression of B19 virus immunopathology or that the rash illness is modified in some way by the frequent use of transfusion.[17] However no rashes were observed in eight Jamican patients recovering without transfusion.

V. MANAGEMENT OF THE APLASTIC CRISIS

The natural history of the aplastic crisis is predictable and the prognosis uniformly good

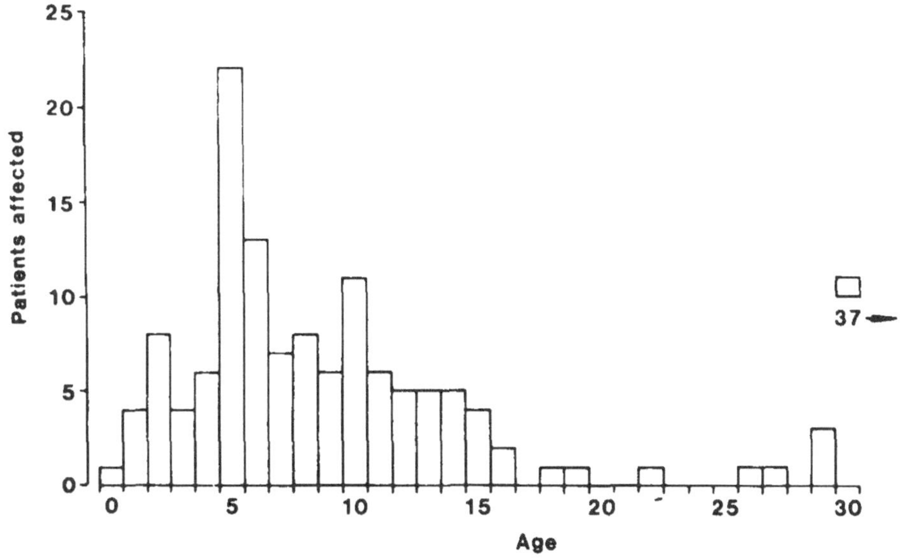

FIGURE 2. Age distribution of 126 aplastic crises among Jamaican SS patients attending the noncohort sickle cell clinics. Inclusion of cohort data would necessarily bias this distribution towards younger ages since all cohort children are currently aged below 13 years.

if the oxygen carrying capacity of the blood can be maintained. The period of aplasia is usually 7 to 10 days and by the time of clinical presentation erythropoiesis has either recommenced or is due to recommence shortly. The greatest danger is in patients with the shortest red cell survival or a longer period of aplasia when the hemoglobin may reach life-threatening levels.

Treatment requires replacement transfusion and because of the predictability of the clinical course, this can usually be done as an outpatient with ambulatory follow-up to ensure that the expected reticulocytosis occurs. Transfusion may modify the degree of reticulocytosis, although this frequently reaches 15 to 20% even in the presence of transfused blood. If lack of iron or folic acid limit the marrow response during the recovery phase of the aplastic crisis, supplementation may be necessary.

In Jamaica, hospital admission is only considered necessary when there are complications such as cardiac failure. Indeed, hospital admission may be contraindicated since infective cases may transmit the disease to susceptible patients in the hospital.[19]

Since almost all cases of aplastic crisis occur in epidemics it is vitally important that medical staff involved in the management of patients with sickle cell disease be informed of the existence of an epidemic. This will increase clinical awareness of the condition and reduce the chance of misdiagnosis. Similarly, families where more than one child is at risk should be informed of the high chance of children being affected simultaneously or within a short period.

VI. EPIDEMIOLOGY OF THE APLASTIC CRISIS

The aplastic crisis in SS disease in Jamaica affects predominantly children. Approximately 90% of cases occur before the age of 15 years (Figure 2), although cases attributable to B19 virus infection have been observed as late as 38 years. This pattern may be modified elsewhere, data from the U.S. suggest that it may affect an older population.[11]

An epidemic pattern is also clear from Jamaican data (Figure 3), epidemics being separated by intervals of 2 to 3 years when no cases are encountered. In the cohort study which

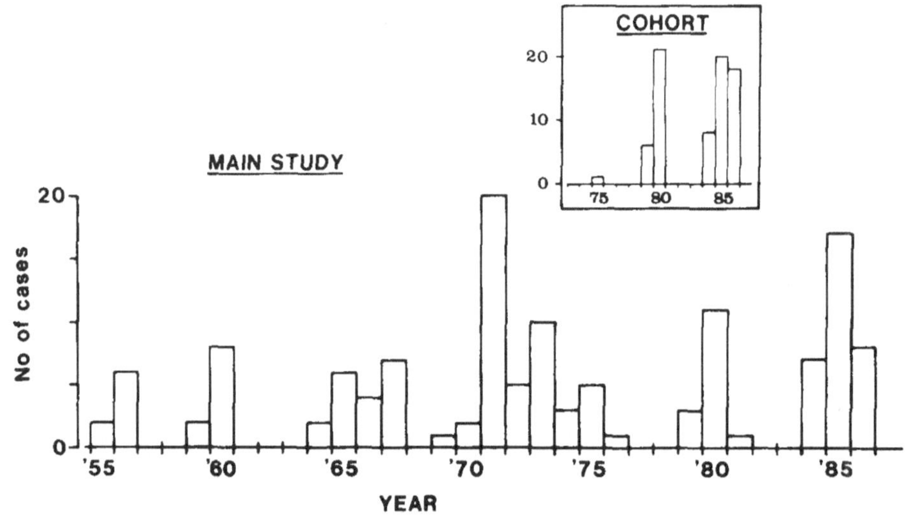

FIGURE 3. Secular occurrence of 131 aplastic crisis among Jamaican SS patients attending the noncohort sickle cell clinics. Data from 74 aplastic crises in the cohort study (commenced in 1973) are inset.

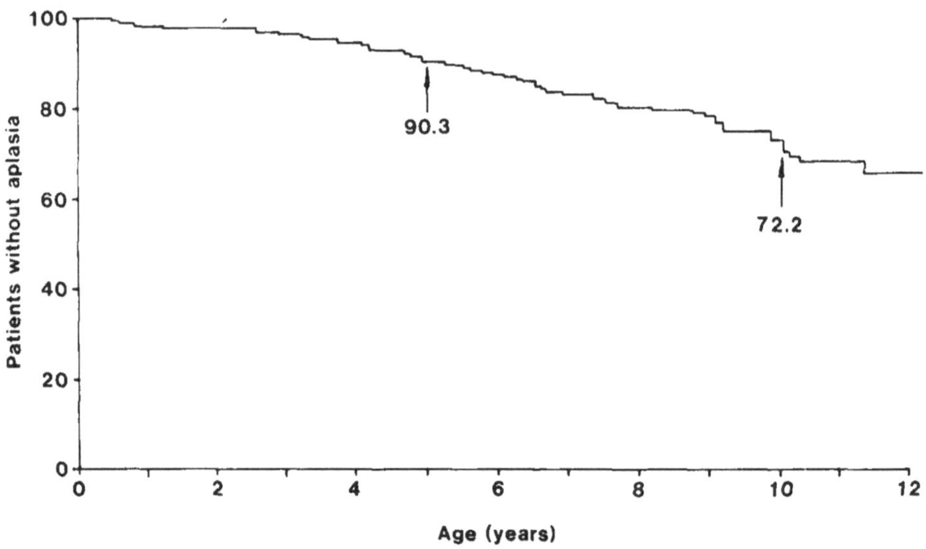

FIGURE 4. Age incidence of aplastic crises in the Jamaican cohort study.

contains a representative sample of 308 children with SS disease followed from birth, it is possible to obtain age incidence data (Figure 4). The youngest affected child was aged 5 months 17 days, 4 (5%) of 74 children were affected before 1 year of age, and only 1 child in the second year. From the age of 2 years there is a relatively steady incidence without any age group demonstrating particular susceptibility, 10% of children having been affected by the age of 5 years, and 28% by the age of 10 years. However this pattern is heavily influenced by the age structure of the cohort study population at the time of epidemics, as is evident when the data are reanalyzed by secular incidence (Figure 5).

The periods of high transmission occurring during epidemics are also reflected in familial involvement and Jamaican data indicate that if an aplastic crisis occurs in a household with other susceptible children aged under 15 years, there is an approximately 50% chance of

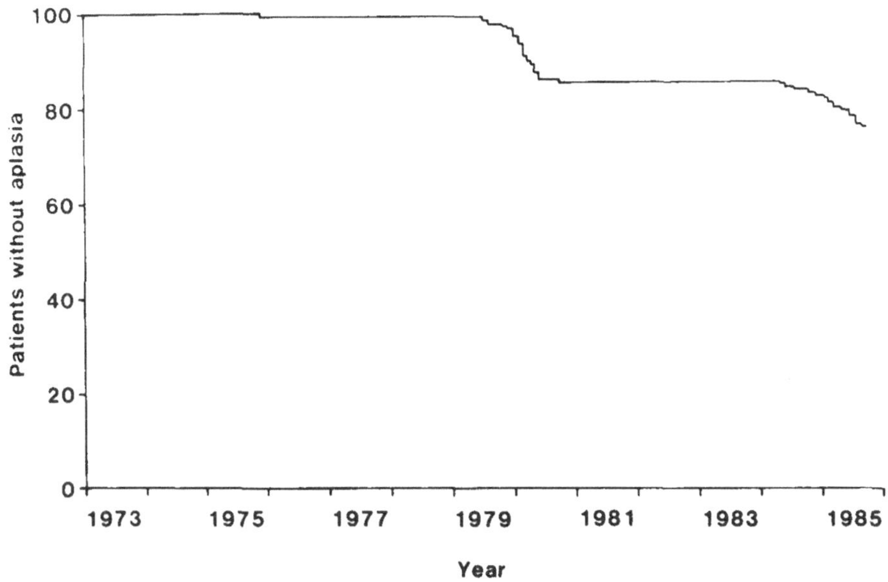

FIGURE 5. Secular incidence of aplastic crises in the Jamaican cohort study.

that child being affected within 3 weeks. The interval between aplastic crises in siblings varies from 0 to 32 days but shows three different patterns. In six sibling pairs, the onset of aplasia was simultaneous or separated by 1 day suggesting that both siblings had been infected from a common source. In seven sibling pairs there was an interval of 6 to 11 days consistent with the latent period observed in experimental infection and suggesting that one had contracted infection from the other. In two pairs, the intervals were longer (20 and 32 days).

Aplastic crises are rarely, if ever, recurrent. When clearly defined to distinguish hypoplasia from aplasia, no patient in the Jamaican clinic has experienced two attacks of aplasia, and neither is there a well-substantiated case in the world literature. In only one report were two episodes described in a patient (Case 1B, Leikin[3]) and review of this case suggests that the second episode was probably infection related hypoplasia. There appears to be convincing evidence that an episode of aplastic crisis confers protection against further attacks.

VII. THE ROLE OF B19 VIRUS

Assessment of the role of the parvovirus in the aplastic crisis has only been possible since awareness of the association and the development of suitable diagnostic techniques. In the last two Jamaican epidemics (1979 to 1981, 1984 to 1986), evidence of recent parvovirus infection has occurred in 108 (96%) of 112 cases. No evidence of parvovirus infection occurred in four cases with apparently typical aplasia so although parvovirus infection accounts for the great majority of Jamaican cases, it is possible that aplastic crises are attributable to other causes.

The corollary of whether parvovirus infection always results in aplastic crisis in susceptible individuals is unclear. There have been four Jamaican patients in whom parvovirus infection was not associated with typical aplasia.[9] Two patients had high HbF levels and so presumed long red cell survival preventing the manifestations of aplasia, but in the other two there was no explanation. One of these had viral DNA in a serum sample taken on admission to hospital with pneumonia but without any signs of aplasia. The frequency with which B19 virus infection occurs without manifestations of aplasia can only be ascertained by a longitudinal serological study identifying the point of seroconversion in each child and these studies are currently underway.

The close association of parvovirus infection with the aplastic crisis and the fact that such aplasias are rarely, if ever, recurrent, suggests that a parvovirus vaccine may be effective in preventing this manifestation. The development of such a vaccine should have a high priority at this time.

REFERENCES

1. **Singer, K., Motulsky, A. G., and Wile, S. A.,** Aplastic crisis in sickle cell anemia. A study of its mechanism and its relationship to other types of hemolytic crises, *J. Lab. Clin. Med.*, 35, 721, 1950.
2. **Chernoff, A. I. and Josephson, A. M.,** Acute erythroblastopenia in sickle-cell anemia and infectious mononucleosis, *Am. J. Dis. Child.*, 82, 310, 1951.
3. **Leikin, L. S.,** The aplastic crisis of sickle cell disease. Occurrence in several members of families within a short period of time, *Am. J. Dis. Child.*, 93, 128, 1957.
4. **Miesch, D., Bàxter, R., and Levin, W. C.,** Acute erythroblastopenia. Pathogenesis, manifestations, and management, *Arch. Intern. Med.*, 99, 461, 1957.
5. **Hilkovitz, G.,** Sickle-cell disease. The "aplastic crisis" and erythroid maturation defect occurring simultaneously in three members of a family, *Arch. Intern. Med.*, 105, 76, 1960.
6. **MacIver, J. E. and Parker-Williams, E. J.,** The aplastic crisis in sickle cell anaemia, *Lancet*, 1, 1086, 1961.
7. **Charney, E. and Miller, G.,** Reticulocytopenia in sickle cell disease. Aplastic episodes in the course of sickle cell disease in children, *Am. J. Dis. Child.*, 107, 450, 1964.
8. **Pattison, J. R., Jones, S. E., Hodgson, J., Davis, L. R., White, J. M., Stroud, C. E., and Murtaza, L.,** Parvovirus infections and hypoplastic crises in sickle cell anaemia, *Lancet*, 1, 664, 1981.
9. **Serjeant, G. R., Topley, J. M., Mason, K., Serjeant, B. E., Pattison, J. R., Jones, S. E., and Mohammed, R.,** Outbreak of aplastic crises in sickle cell anaemia associated with parvovirus-like agent, *Lancet*, 2, 595, 1981.
10. **Kelleher, J. F., Mortimer, P. P., and Kanimura, T.,** Human serum "parvovirus" a specific cause of aplastic crisis in children with hereditary spherocytosis, *J. Pediatr.*, 102, 720, 1983.
11. **Rao, K. R. P., Patel, A. R., Anderson, M. J., Hodgson, J., Jones, S. E., and Pattison, J. R.,** Infection with parvovirus-like virus and aplastic crisis in chronic hemolytic anemia, *Ann. Intern. Med.*, 98, 930, 1983.
12. **Duncan, J. R., Potter, C. G., Cappellini, M. D., Kurtz, J. B., Anderson, M. J., and Weatherall, D. J.,** Aplastic crisis due to parvovirus infection in pyruvate kinase deficiency, *Lancet*, 2, 14, 1983.
13. **Goldstein, A., Anderson, M. J., and Serjeant, G. R.,** Clinical features of parvovirus associated aplastic crisis in children with homozygous sickle cell disease, *Arch. Dis. Children*, 62, 585, 1987.
14. **Okcuoglu, A.,** Erythropoietin in acute erythroblastopenia of sickle cell anemia, *Turk. J. Pediatr.*, 6, 225, 1964.
15. **Hammond, D., Shore, N., and Movassaghi, N.,** Production, utilization and excretion of erythropoietin. I. Chronic anemias. II. Aplastic crisis. III. Erythropoietic effects of normal plasma, *Ann. N.Y. Acad. Sci.*, 149, 516, 1968.
16. **Haddy, T. B., Lusher, J. M., Hendrickes, S., and Trosko, B. K.,** Erythropoiesis in sickle cell anaemia during acute infections and crisis, *Scand. J. Haematol.*, 22, 289, 1979.
17. **Anderson, M. J., Davis, L. R., Hodgson, J., Jones, S. E., Murtaza, L., Pattison, J. R., Stroud, C. E., and White, J. M.,** Occurrence of infection with a parvovirus-like agent in children with sickle cell anaemia during a two-year period, *J. Clin. Pathol.*, 35, 744, 1982.
18. **Anderson, M. J., Higgins, P. G., Davis, L. R., Williams, J. S., Jones, S. E., Kidd, I. M., Pattison, J. R., and Tyrrell, D. A. J.,** Experimental parvoviral infection in humans, *J. Infect. Dis.*, 152, 257, 1985.
19. **Evans, J. P. M., Rossiter, M. A., Kumaran, T. O., and Marsh, G. W.,** Human parvovirus aplasia: case due to cross infection in a ward, *Br. Med. J.*, 288, 681, 1984.

Chapter 6

RASH ILLNESS DUE TO B19 VIRUS

M. J. Anderson

TABLE OF CONTENTS

I. INTRODUCTION

When B19 virus was first discovered[1] the majority of individuals in whom the viremia was detected were healthy blood donors. Follow-up of a number of these donors revealed a variety of minor symptoms.[2,3] In addition, one had a rubelliform rash 4 days after the viremia and four children of one of the donors had itchy rashes 3 to 5 weeks after their mother had donated blood. Some years later, further blood donor screening in the author's laboratory detected viremia in two donors who were husband and wife. Both adults had minor, nonspecific symptoms but two of their children developed an erythematous rash illness 3 weeks after the mother was viremic.[3a] On reflection, it seemed that an erythematous rash illness was the only specific illness in otherwise healthy individuals that might represent the common clinical manifestation of B19 virus infection. Bearing in mind that B19 infection was known to be most common in childhood, consideration was given to those rash-illnesses of childhood of unknown, but presumed viral, etiology. Two such diseases are erythema infectiosum (fifth disease) and roseola infantum (sixth disease). At the time of writing the latter remains of unknown etiology, but there have been extensive investigations of erythema infectiosum and B19 infection.

Tschamer[4] is usually credited with the first accurate description of erythema infectiosum, although he regarded it as a variant of rubella. During the next few years it was recognized as a distinct disease, and in 1899, Sticker[5] first used the name "erythema infectiosum". There was a fashion at about this time for numbering the distinct childhood exanthemata. Accordingly the term *fifth rash disease,* now abbreviated to *fifth disease,* was applied by Cheinisse.[6] These two names, erythema infectiosum and fifth disease, together with "slapped-cheek disease" are the names most commonly given to an old established disease which now needs some redefinition with the advent of specific tests for B19 virus infection. This chapter reviews the clinical and virological aspects of erythema infectiosum (EI).

II. THE CLASSICAL PICTURE OF EI

There have been many descriptions of outbreaks and epidemics of EI throughout the world. These have been reviewed and analyzed by Cherry and readers are referred to Cherry's textbook description and extensive references.[7]

The incubation period of EI has been variously described as ranging from 4 to 14 days. (It is now clear from volunteer studies (see Chapter 7) that the true incubation period is 17 to 19 days. However virus excretion occurs 6 to 10 days after infection with B19 virus so the case-to-case interval is of this order.) Prodromal symptoms (low-grade fever, headache, and malaise) have been noted in only a few patients 1 or 2 days before the appearance of the rash. Classically the rash first appears on the face. Erythema of the cheeks and relative circumoral pallor contributes a sharply demarcated "slapped-cheek" appearance. A discrete, erythematous maculo-papular appearance develops on the limbs and trunk 1 to 4 days later. The rash has a tendency to become more confluent with central clearing giving a lacy or reticular appearance which is regarded as very characteristic of EI. Finally, the rash has a tendency to alternately clear and then recrudesce for a period of about 3 weeks. Recrudescence is often related to changes in surface body temperature brought on by exercise, bathing, or sunlight. The rash is often associated with pruritis, especially in adults, is more common on extensor surface, and is only rarely seen on the palms and soles. In general no enanthem is found in cases of EI.

Many variations of this classical picture have been described, but in the absence of specific diagnostic tests it is possible that extreme variations in the rash are due to other causes. Associated symptoms such as fever, headache, respiratory tract and abdominal symptoms are mild and only occur in 10 to 20% of cases in children although they are rather more common in adults. The only significant and frequent complication of EI is joint involvement. Pain, stiffness, and swelling of the joints is found in only a minority (about 10%) of cases in children but in a majority of cases (about 70%) of adults, especially women. There are case reports of hemolytic anemia and encephalitis in association with EI, but in the vast majority of cases it is a mild, short-lived disease with an excellent prognosis. Rubella, scarlet fever, other viral exanthems, hypersensitivity, and other allergic reactions are acknowledged to be problems in the differential clinical diagnosis.

Epidemics of EI most often occur in the winter and spring. After analyzing 23 epidemics Cherry[7] states that 19 had their onset in the period December to May and 17 of them peaked in March, April, or May. There is also a cyclic pattern with peaks of disease activity every 6 years with the peak periods lasting up to 3 years. The attack rate is highest in children aged 5 to 14 years and because of this outbreaks are often centered on elementary schools in which there may be attack rates of 25% or greater.

III. EARLY INVESTIGATIONS INTO THE CAUSATIVE AGENT OF EI

The first study to isolate the causative agent of EI was conducted with material from cases occurring as an outbreak among children in Florence in 1945.[9] The illness was characterized by fever, a benign, short-lived rash, and lymphadenopathy was frequently noted. The rash was said to be like measles or scarlet fever but a past history of scarlet fever or measles was not associated with protection. Plasma and nasopharyngeal secretions were inoculated intraallantoically into 9-day-old fertile hen's eggs. After 3 to 4 days incubation at 37°C the chorioallantoic membrane was edematous with small bleeds away from the inoculation site and whitish nodules in the perivascular areas of the membrane. The same effects were observed after two or three passages and were neutralized by convalescent sera taken from patients 20 or 30 days after the rash. No follow-up studies of these original "isolates" or attempts to repeat the work with material from other cases of EI have been reported. Most recently B19 virus has been inoculated into fertile hen eggs, but no evidence of replication has been found.[9a]

Another possible etiological agent was isolated during investigation of an outbreak of EI in Pennsylvania in 1956.[9] Heparinized blood, nasopharyngeal samples, and stool specimens were inoculated onto cultures of rhesus monkey kidney which were observed for cytopathic effect (CPE) and passed blind to fresh cultures if no CPE was observed. Evidence of an infectious agent appeared in the second passage of one stool specimen, the third passage of another stool specimen, and the fourth passage of a throat washing. The CPE observed consisted of the appearance of multinucleated giant cells, with vacuolated cytoplasm, and inclusions in both the nuclei and cytoplasm. The agent passed with ease and CPE developed from day 3 after inoculation. A fourfold or greater rise in antibody titer was found in five of nine pairs of sera from cases using a concentrated cell and culture fluid antigen in a complement fixation test. The original report also draws attention to two similar isolates from another outbreak in Pennsylvania. The authors drew attention to two problems in judging the significance of the isolates. First they noted the problems of latent viruses in monkey cell cultures but clearly went to some lengths to avoid obviously contaminated cultures. They also note the similarity between the CPE observed and that produced by measles virus. There appears to be no follow-up work on these isolates and unfortunately they are no longer available for study.

Balfour and colleagues conducted an extensive virological investigation on a large number of patients and contacts presenting during a 19-month outbreak of rash illness between January 1969 and July 1970.[10] The rash was considered to be EI because of the characteristic progression of the rash (initial involvement of the cheeks, followed by spread to involve trunk and limbs, and finally the development of a reticular appearance) and its long duration (mean 11 days, range 2 to 39 days). Pharyngeal and anal swabs were obtained from 102 patients and 200 family contacts. Paired serum specimens were obtained from 68 patients. Swabs were inoculated into a range of cell cultures which were examined for cytopathic effect, hemadsorbtion, and interference and passed three times before being considered negative. Rubella virus was isolated from eight patients and two asymptomatic siblings of these patients. Paired sera were available from six of the patients and in all six a significant rise in rubella HI antibody titer was found. Diagnostic rising titers were also found in two other patients from whom rubella virus was not isolated and thus there were 10 of 102 patients with evidence of recent rubella infection. Inoculation of one of the isolates into five male volunteers produced an exanthem which began with slapped cheeks in four and lasted on average 6 days. The rash did not develop reticular appearance.

Subsequent work[11,12] on cases of EI failed to provide any evidence for rubella virus being the cause of EI. No virus could be isolated from patient material and a history of rubella immunization did not confer protection against EI.[12] It seems clear now that there was a cluster of cases of rubella among the original group studied by Balfour and colleagues. All the rubella cases occurred during an 11-week period whereas there were two peaks of exanthematous illness as a whole, one at the same time as the rubella cases and one approximately 12 months later. All the rubella cases had lymphadenopathy whereas it was present in only 34% of the virus-negative patients. These features are in accord with recent observations that B19 virus rash-illness often mimic rubella but that lymphadenopathy is more a feature of the latter.

In the 102 patients investigated by Balfour, echovirus 12 was found in four (three of them siblings). There was serological evidence of recent infection in only one, and these findings probably represent either an incidental finding or a small number of echovirus rashes in the group investigated. Thus after several extensive investigations using the classical virological techniques of the time, no causative virus for EI was found. This picture changed with the advent of specific tests for B19 virus infection in the early 1980s.

IV. THE FIRST OUTBREAK OF EI INVESTIGATED FOR B19 VIRUS INFECTION[13]

In May of 1983, a case of scarlet fever was diagnosed clinically in a school in London. Advice was sought from the Communicable Diseases Surveillance Centre, London, about possible closure of the school. Inquiries revealed that there had been a number of recent similar cases, resembling EI more than scarlet fever, and it appeared that an outbreak of EI was in progress. A questionnaire was sent out to the parents of all 430 children in the school inquiring about the details of rash illnesses during the preceding 3 months. Questionnaires were returned for 369 of the children and 162 (44%) had suffered a rash illness. This most commonly involved the cheeks (87%), and arms and legs (78%), and was less frequently seen on the trunk. The rash was described as lacy in 60%, was noted to recrudesce in 67%, and was associated with itchiness in 46% of cases. Associated symptoms were mild with fever (27%), headache (28%), and sore throat (32%) being the most common. Joint involvement was noted in only 12 (7%) of the cases in the children. The first case occurred on February 28, 1983 and the last case occurred on June 4, 1983; 75% of cases occurred during April and the first half of May (Figure 1). The ages of the children in the school ranged from 4 to 11 years and the rash illness occurred most commonly among the 7- to 10-year-olds. In all clinical and epidemiological respects this outbreak was very similar to the classical descriptions of EI.

Permission was granted to take blood specimens from 30 children involved in the outbreak and 23 members of eight families with an affected child. Amongst these eight families there were a further six cases of rash-illness (three parents, three siblings). Thus there were 36 cases and 17 well family contacts investigated for evidence of recent B19 virus infection.

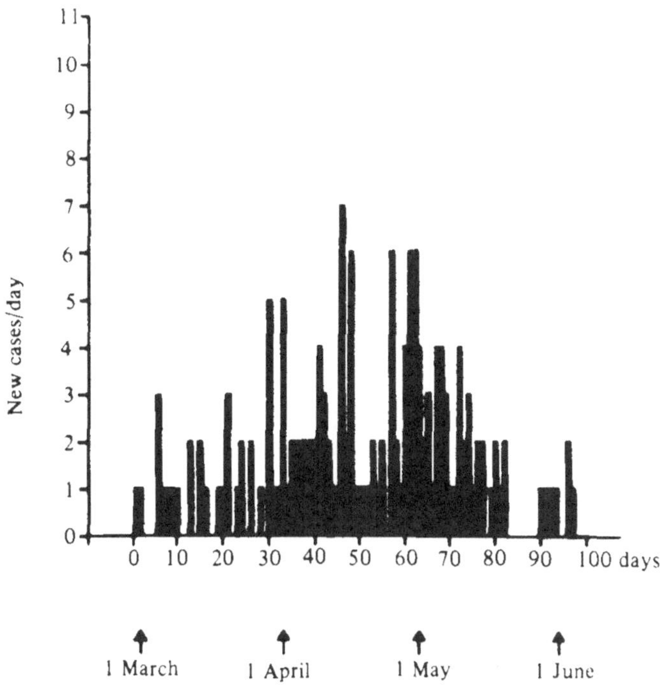

FIGURE 1. Distribution of 162 cases of erythema infectiosum by date of onset of rash. (From Anderson, M. J., Lewis, E., Kidd, I. M., Hall, S. M., and Cohen, B. J., *J. Hyg.*, 93, 85, 1984. Reproduced by kind permission of the editors of *The Journal of Hygiene*.)

In addition, 22 sera from well children from another part of London and from ten heterosexual couples of similar ages to the parents of the affected children were included as controls.

Each serum specimen was tested for B19 virus DNA by a dot hybridization technique and for parvovirus antigen by radioimmunoassay (RIA). None of the samples from the 36 cases of EI were positive for viral DNA or antigen. However the sera from all 36 cases contained B19 virus-specific IgM with an overall trend for the amounts of IgM to be lower with increasing time between the onset of the rash and blood sampling (Figure 2). In contrast, none of the sera from well family contacts, or adult controls contained this class of specific antibody; one serum from the 22 control children gave a weak reaction (equivalent to one unit per milliliter), all others were negative. All sera from the cases of EI contained B19 virus-specific IgG antibody as did all four sera from the children and 12 of the 13 adults who were family contacts of cases but who remained well. Of the controls 4 (18%) of 22 children and 16 (84%) of 19 adults had IgG antibody; these rates of positivity being as expected for the age groups concerned.

The serological investigations outlined above clearly indicated that this epidemic of EI was associated with B19 virus infection. Cases had specific IgM and IgG antibody indicating recent infection and well contacts had specific IgG only indicating infection in the remote past. Since rubella had been implicated as a possible cause of EI in the past all sera from cases were tested for rubella-specific IgM. In 6 of the 36 cases trace amounts of this antibody were detected, which could have been compatible with rubella 2 to 4 months previously. The onset of the rash was 4 to 60 days prior to the serum sample being taken and thus there was no correlation between the occurrence of EI and rubella infection in these cases. The finding clearly represents the phenomenon of heterologous reactions found in IgM antibody, capture RIAs in rubella, and B19 virus infections.[14]

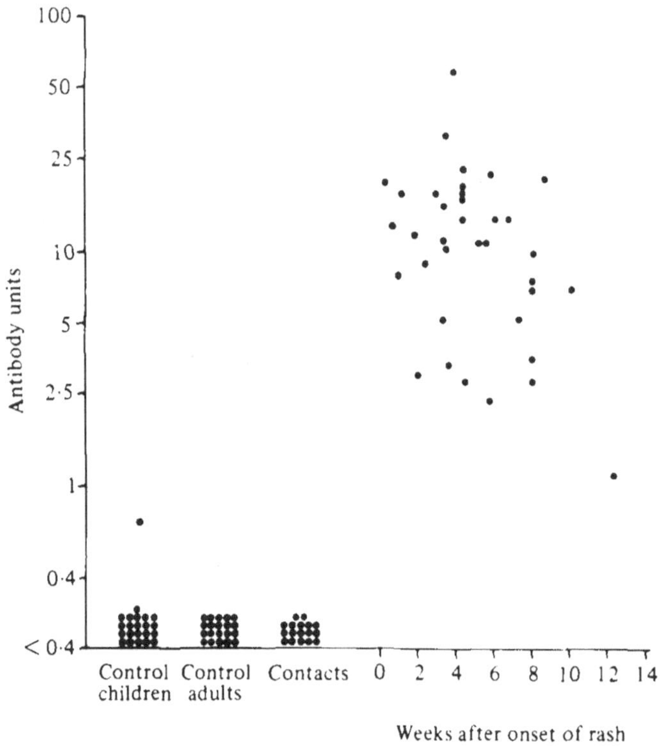

FIGURE 2. Parvovirus-specific IgM antibodies in controls, asymptomatic contacts, and cases of erythema infectiosum. (From Anderson, M. J., Lewis, E., Kidd, I. M., Hall, S. M., and Cohen, B. J., *J. Hyg.*, 93, 85, 1984. Reproduced by kind permission of the editors of *The Journal of Hygiene*.)

No parvovirus antigen or DNA was found in the sera from the 36 cases of EI. These sera were taken relatively late after the start of the illness and only four were taken within the first week. Further it was argued from the beginning that the rash was likely to be a rather late event in infection and that its properties were consistent with the requirement for an immunological component as in the exanthems of measles and rubella. Volunteer studies (see Chapter 7) have confirmed these ideas and therefore diagnosis of established rash-illness due to B19 virus infection must depend upon serology.

V. FURTHER EVIDENCE FOR THE ASSOCIATION OF EI AND B19 VIRUS INFECTION

The link between specific virus and disease was rapidly confirmed by the retrospective study of cases of EI in Japan.[15] The disease occurred in Tokyo in 1974, 1975, and 1977 and was widespread in Japan in 1980 and 1981. Sera from 17 cases in a school outbreak in 1977 and 11 cases from the widespread outbreak in 1981 were investigated for evidence of recent B19 virus infections. All cases were children age 5 to 11 years and 17 sera from age-matched control children were also tested. Specific B19 virus IgM was found in all 17 of the children from the 1977 outbreak and in 10 of the 11 from the 1981 epidemic. In those from whom paired sera were available, the amount of specific IgM fell during convalescence and amounts of specific IgG rose. None of the control sera had specific IgM antibody and ten (59%) had IgG antibody. The one case which was specific IgM antibody negative was also lacking specific IgG antibody and virus antigen was not detected in the serum.

Investigations of further cases from the Japanese epidemic in 1981 were reported.[16] In the month of January, 676 cases were reported from children's clinics in Sendai. Paired sera were obtained from 19 of these, the acute specimen being taken 1 to 13 days after onset, and the convalescent sample 18 to 31 days after onset. At least one of the sera from all 19 cases contained B19 virus IgM antibody and in 17 cases both sera contained such antibody. There was no evidence of measles infection in any of the patients but there was evidence of concurrent rubella virus infection in one case. Thus, in addition to the London outbreak of 1983 it is clear that cases of EI in Japan in 1977 and 1982 were due to B19 virus infection.

VI. A PROSPECTIVE STUDY IN WINNIPEG[17]

All the studies described above were retrospective in nature and there was no opportunity to determine the full range of effects of B19 virus infection circulating in communities or families. However, in the spring of 1980, outbreaks of exanthematous illness were noted in several communities in Manitoba, Canada. Later in the spring of the same year the occurrence of exanthematous illness among school children in Winnipeg provided the opportunity to study such illnesses prospectively. The subjects for the study were from four schools in Winnipeg and consenting families were enrolled into the study if the index case in the family had been ill for less than 48 hr. Seventeen families were enrolled into the study but five were later excluded because the index case proved to be measles (one case) or rubella (four cases). The index cases in the remaining 12 families attended schools in which no measles or rubella infections were detected. There were 13 primary rash illnesses in these families (two siblings in one family had simultaneous illnesses) and six secondary cases in family contacts (three children, three adults). The rash illness most closely resembled EI but there were some atypical features. Respiratory and systemic symptoms were more prominent than is usual. The rash occurred on the palms and soles in a third of the patients and the typical slapped cheek and lacy reticular rash were present in only one of the patients. In addition to the rash illness in contacts, mild systemic and respiratory illness without a rash occurred in three children and four adult family contacts.

At the time of the outbreak extensive virological investigations excluded infection with measles, rubella, enteroviruses, and other known causes of exanthemata. Sera from 9 of the 13 primary index cases were available for B19 virus serology and specific IgM antibody was detected in eight. Sera from three of the six secondary cases with a rash were also tested and all three had specific IgM antibody. In addition, three of six family contacts who developed a nonrash illness were positive as were 2 of 19 well family contacts. By contrast none of the serum samples from 28 cases of measles and rubella during the study period contained B19 virus-specific IgM. In addition to the serological evidence of recent infection there was evidence that B19 virus was circulating at the time. Two serum specimens contained viral DNA. One was obtained at the first family visit from the mother of a child with rash illness. The mother had prodromal symptoms at the time, a rash developed 1 day later, and a second serum specimen showed the presence of B19 virus-specific IgM and IgG. The second DNA positive serum was obtained from a child who had two siblings with rashes and one with a nonrash illness all of whom were B19 virus-specific IgM positive. The specimen containing virus was obtained at the final family visit and there was no follow-up information.

The specimens which had been collected for virus isolation were also tested for B19 virus DNA and one throat swab yielded a positive result. This was from a teacher with a nonrash illness who was excreting herpes virus from the throat at the same time. Acute and convalescent sera had just detectable amounts of specific IgM but substantial amounts of specific IgG. This pattern of results raises intriguing possibilities. They suggest that reinfection might have occurred in this individual or it is just possible that latent B19 infection was reactivated by concurrent herpes simplex infection. In any case the recovery of virus from the throat swab suggested that transmission was via the respiratory route, a notion that was later confirmed in volunteer studies (see Chapter 7).

VII. ENDEMIC RASH ILLNESS DUE TO B19 VIRUS INFECTION

The atypical features of the Winnipeg outbreak described above suggested that B19 virus was capable of causing rash illness other than classical EI. The studies also confirmed that asymptomatic infections and illnesses without a rash can occur. The application of serological tests for B19 virus infection during the past 2 years in the U.K. has shown that sporadic cases and family clusters of B19 virus associated rash illness occur throughout the year. A total of 31 such cases were diagnosed during 1983 in the U.K.[27] Sixteen of these occurred in seven family clusters, three in a residential home, and the remainder were individual cases. One or more of these cases occurred in each month of the year except January, August, and September. Furthermore although 21 of the cases were diagnosed clinically as EI, the remaining 10 were diagnosed as rubella-like illness. It is now clear that the differential diagnosis of erythematous rashes due to B19 virus infection on clinical grounds alone is not possible except during outbreaks of classical EI with slapped cheeks and lacy, reticular rashes on the limbs. Rubella is often given as the diagnosis and the other common clinical diagnoses for B19 virus infection are allergic rash or unspecified viral rash. During a 15-month period (January 1983 to March 1984) 232 sera from 164 patients with rubelliform rash illness were submitted to a single diagnostic laboratory in London. All were tested for evidence of both recent rubella and B19 virus infection.[18] Rubella infection was confirmed in 39 (24%) and B19 virus infection in 13 (8%). No dual infections were found. Lymphadenopathy was seen significantly ($P < 0.5$) more often in rubella cases, but joint involvement occurred in 18% of rubella infections and in more than twice that proportion of B19 virus infections. The seasonal incidence of the two infections (peak numbers in May to August) was the same. Again there is no doubt that rubella and B19 virus can cause similar illnesses and, although the presence of lymphadenopathy may favor a diagnosis of rubella, specific laboratory tests are the sole means of determining accurately the viral cause.

B19 virus infection is common; at least 60% of adult populations have evidence of past infection. It is world-wide in distribution. The common manifestation of infection is an erythematous rash and such illnesses have been shown to be due to B19 virus infection in the U.K., France, Holland, Germany, Sweden, Norway, Iceland, Japan, Canada, Italy, Finland, and the U.S. These cases have occurred at various times from 1962 to the present day. Overall B19 virus infection may be asymptomatic, may cause a mild "flu-like" illness, or may be associated with sporadic cases, family clusters, or larger outbreaks of erythematous rash illness. These latter are often labeled EI and, to date, over 95% of cases of EI which have been appropriately investigated have been shown to be due to B19 virus infection.

VIII. PURPURA AND B19 VIRUS INFECTIONS

An erythematous rash is not the sole skin manifestation of B19 virus infection. In a small number of individuals easy bruising may be noted by the patient and a purpuric rash is evident on examination. Four such cases have been described from France.[19] An 11-year-old boy presented with purpura at the same time as his mother and sister presented with typical EI. An 11-year-old girl with β-thalassemia presented with purpura at the same time as suffering an aplastic crisis. Two other patients (a 5-year-old girl and a 33-year-old woman) presented with a similar non-necrotic petechial purpura. All had prodromal symptoms of fever, shivers, and rhinopharyngitis for 48 hr prior to presentation. Purpura was present on the legs of all patients; the thorax and upper limbs were involved in two and the abdomen in another. The 5-year-old girl was diagnosed as having Henoch-Schonlein purpura. B19 virus was detected in sera taken on the days of admission from the woman and the boy and both subsequently seroconverted. The first blood from the patient with β-thalassemia was taken 4 days after the onset of purpura and 5 days after the onset of the aplastic crisis. Virus was not found but the presence of B19 virus-specific IgM was diagnostic of recent infection. The 5-year-old girl with Henoch-Schonlein purpura also had specific IgM. Platelet counts were greater than $150 \times 10^9/\ell$ in all patients and the purpura resolved within a few days.

Two other cases have been reported[20] in which recent B19 virus infection in association with purpura was diagnosed by the presence of high concentrations of B19 virus IgM antibody. A 5-year-old girl was noticed to have bruising and developed red cheeks and a purpuric rash on the thighs a few days later. The patient was thrombocytopenic ($33 \times 6^9/\ell$) when first investigated but the platelet count rapidly returned to normal. The other case was a 50-year-old man with a purpuric rash on his body and legs. This case is notable because there was still some persistence of the rash 22 days later.

The author's experience is similar to the above. The occasional patient with recent B19 infection will present with purpura. Some may have a low platelet count at the time but most are not thrombocytopenic. The pathogenesis of the purpura remains obscure although volunteer studies (see chapter 7) show that there is a fall in platelets commencing at the time of the viremia. This explains the finding of virus in two of the patients described above.

The author has also diagnosed recent B19 infection in a case of Henoch-Schonlein purpura. No formal studies of B19 virus serology in collections of cases of this condition have been reported so it remains possible that the virus is a cause of at least some cases. Cases of idiopathic thrombocytopenic purpura have, however, been studied[20a] and B19 virus has not shown to be a cause of this condition.

IX. JOINT INVOLVEMENT IN B19 VIRUS INFECTION

Joint symptoms have been recognized as a complication of EI since early descriptions of outbreaks of the disease. In the large outbreak documented by Ager et al.,[21] joint involvement was noted in 8% of the 307 cases in children and 77% of the 57 adult cases. In the first London outbreak shown to be due to B19 virus 7% of the 162 children and 81% of the 16 adults had joint symptoms.[13] The complication occurs most frequently in adult women. This pattern of the frequency of complicating arthropathy in cases of erythematous rash illness due to B19 virus infection has been consistently found with the application of specific diagnositic tests over the last 2 years. Two large studies of the relationship between B19 virus and arthropathy have been reported.[22,23] During a large outbreak of EI in a region of Scotland the occurrence of joint symptoms in some cases prompted a more detailed study.[22] Patients were selected for study not on the basis of the presence of an erythematous rash but on the basis of presentation with joint pains. There were 42 patients in the study 32 (76%) were adult women (age range 16 to 56 years), 5 (12%) were men (aged 19 to 59 years), and 5 (12%) were children aged 6 to 11 years. Detailed clinical information was available for 30 of the patients (27 adults, 3 children) from visits to a rheumatology clinic (17 patients) or from a questionnaire sent to their primary care physicians (13 patients). All three children had a rash and typical "slapped cheeks" were present in two. Only 13 of the 27 adults had a rash before the joint symptoms; it was erythematous, mildly itchy, and described by a medical attendant as rubelliform. In seven there were no prodromal features at all. All the adults except one had a symmetrical arthritis of the small joints of the hand. Of those for whom complete information was available the joints most commonly affected were the proximal interphalangeal joints followed by the metacarpophalangeal joints, knees, wrists, and ankles. The joint pains were accompanied by stiffness and all those examined had swelling of the finger joints. The arthritis was generally mild but some cases were absent from work. The arthritis resolved in 2 weeks in 70% and had settled within 4 weeks in all but two in whom it lasted $6^1/_2$ and 24 weeks. In the children the arthritis was less symmetrical, was accompanied by swelling, and lasted $2^1/_2$, 13, and more than 28 weeks. In this last child the arthritis commenced 10 days prior to the rash whereas in all the others the interval between rash and arthritis was -1 to $+6$ days. This must raise doubts about whether the rash illness and the arthritis were due to the same cause in the patient with the most persistent arthralgia. In all patients B19 virus infection was diagnosed by the presence of specific IgM antibody at the time of presentation except one who initially had a viremia and then subsequently seroconverted. None of the patients were rheumatoid factor positive.

In a second study cases were drawn from those attending an early synovitis clinic in the west of England.[23] Primary care physicians were invited to refer patients as soon after the onset of joint pains as possible in order to investigate novel diagnostic and prognostic tests to distinguish rheumatoid arthritis from transient reactive inflammatory arthritis. Retrospective study of sera from 153 patients presenting between May 1979 and August 1980 for evidence of B19 virus infection was possible when specific serological tests became available. B19 virus-specific IgM was found in 19 of the 153 patients. Other possible infective causes of arthritis were found in five patients (*N. gonorrhoea*, 2; Epstein-Barr virus infection, 2; and rubella infection, 1). A total of 104 (68%) of the group as a whole were women but all 19 of the recent B19 virus infections were women. Their age range was 18 to 58 years (median 38) in contrast to the group as a whole (range 15 to 88, median 51 years). A total of 13 of the 19 patients with recent B19 virus infection gave a history of illness in the 8 weeks before the onset of joint symptoms. In most the symptoms were nonspecific involving the respiratory and gastrointestinal tract but four had a history of a rubelliform rash. In two patients the rash appeared 7 and 4 weeks before the arthralgia, in one the rash and arthralgia appeared simultaneously, and in the fourth the rash appeared 3 days after the arthralgia. (Again the interval of weeks between rash and arthralgia suggests that both may not have the same etiology.) Of the 19 patients, 18 had moderately severe symmetrical polyarthritis that started in the hands (ten cases) or knees (seven cases), and spread to involve wrists, ankles, feet, elbows, and shoulders. Other joints were occasionally affected and in the 19th patient there was only a history of pain and stiffness of the lower back. In the other patients joint stiffness and variable swelling was noted. Although the symptoms improved within 2 weeks in most patients it cleared completely within this time in only two patients. Some symptoms persisted for more than 2 months in 17 patients and for more than 4 years in 3 patients. Persistence of joint symptoms was not found to be associated with long-term joint damage or functional incapacity. Rheumatoid factor was found only in the patient whose sole symptom was low back pain.

The authors of the above study state that three of these patients could have been diagnosed as an early benign form of rheumatoid arthritis but the almost complete absence of rheumatoid factor and associated laboratory abnormalities led them to conclude that B19 virus was unlikely to be a cause of rheumatoid arthritis. This conclusion was reinforced by results from France which showed that the frequency of B19 virus antibody in 105 sera from patients with seropositive rheumatoid arthritis was not significantly different from that of a healthy population.[24] Three of the sera had specific IgM antibody indicating recent infection but the patients had had rheumatoid arthritis for 1, 9, and 13 years. There is however a report[25] of a husband and wife with recent B19 virus infection. The wife had finger-joint symptoms for a few days with no sequelae. The husband had a symmetrical polyarthritis of hands, wrists, elbows, ankles, knees, and spine which lasted at least 8 months. During the first 6 months the rheumatoid factor titer rose to high levels although ESR, C-reactive protein, and immunoglobulin levels were normal at 2 months. The laboratory results do not indicate classical rheumatoid disease but such a pattern of results could clearly lead to diagnostic confusion.

X. CONCLUSION

It is now clear that B19 virus is the elusive agent of EI. This virus does not replicate in the cell-cultures conventionally used in diagnostic virology and indeed has so far evaded all attempts at culture. It is not surprising therefore that investigations to determine the cause of EI prior to 1983 were unsuccessful. Moreover volunteer studies (Chapter 7) suggest that in the majority of cases the rash and associated symptoms occur 1 week or so after the period when virus excretion is detectable by current methods. The incubation period of B19 virus rash illness seems to be 17 to 18 days. However virus excretion occurs between days 7 to 12 after inoculation and therefore case-to-case intervals in outbreaks of EI are likely to be of this shorter duration.

The discovery of the causative virus of EI was dependent upon the development of serological tests for B19 virus infection. The stimulus to this development was the necessity of being able to diagnose with accuracy the cause of aplasia in chronic hemolytic anemia. The link with EI was established as a result of the logical search for the common manifestation of a common infection. All outbreaks of EI so far investigated have been shown to be due to B19 virus. Most of the stored sera available for analysis have been from classical outbreaks of EI with infection common in children aged 4 to 11 years and with the slapped-cheek and reticular nature of the rash. However the routine application of B19 virus serology by a few laboratories for the last 2 years has shown that rash-illness due to this virus is not confined to classical EI. The cases do not always occur in outbreaks and sporadic cases and family outbreaks constantly occur. Older children and adults are affected and this is to be expected since up to 40% of adult populations have no serological evidence of previous infection. Cases occur throughout the year, but there is nevertheless, a tendency for B19 virus infection to be most common in late winter, spring, and early summer. Clinical observations of EI and aplastic crises (Chapter 5) indicate that there are undoubtedly years of high incidence but specific serological tests for B19 virus infection have not yet been available long enough to define the periodicity.

The clinical features of the rash-illness are variable. Slapped-cheeks are certainly not always present and neither is the reticular appearance of the rash. Both tend to be seen in children but not in adults. Overall rash-illness due to B19 virus infection is as often rubelliform as it is like classical EI. It is interesting that EI was initially thought to be a form of rubella and that the two viruses *can* co-circulate and cause clinically indistinguishable illness. The presence of lymphadenopathy favors a diagnosis of rubella but specific serological tests are the only certain way of making the virological diagnosis.

Joint involvement is the most significant complication of rash-illness due to B19 virus. The precise pathogenesis is unclear; it is likely to be immune mediated, since the joints are affected in most cases after the period of the viremia and at a time when there is detectable specific antibody particularly of the IgM class. In most instances the condition resolves within 2 weeks. However one study,[23] which will have tended to select for the most severe cases, indicates that in some patients it may last for some months or even years. This is rare and the B19 virus arthropathy can be regarded as very similar to rubella arthropathy. B19 virus does not appear to be a candidate cause of rheumatoid arthritis, but the isolation of a different parvovirus from synovial cells of patients with rheumatoid arthritis[26] will undoubtedly stimulate further work in this area.

REFERENCES

1. **Cossart, Y. E., Field, A. M., Cant, B., and Widdows, D.,** Parvovirus-like particles in human sera, Lancet, 1, 72, 1985.
2. **Paver, W. K. and Clarke, S. K. R.,** Comparison of human fecal and serum parvo-like viruses, *J. Clin. Microbiol.*, 4, 67, 1976.
3. **Cant, B. and Widdows, T.,** Data published as an abstract for the Conf. Int. Soc. Blood Transfus., 1975, 73.

3a. **Anderson, M. J.,** unpublished observations.

4. **Tschamer, A.,** Veber ortliche Rotheln, *Jahrb. Kinderheilkd.*, 29, 372, 1889.
5. **Sticker, G.,** Die neue Kindersenche in der Vungebung von Giessen (Erythema Infectiosum), *Z. Prakt. Aerzte*, 40, 121, 1899.
6. **Cheinisse, L.,** Une cinquieme maladie eurptive: Le megalerytheme epidemique, *Sem. Med.*, 25, 205, 1905.
7. **Cherry, J. D.,** Erythema infectiosum in, *Textbook of Pediatric Infectious Diseases*, Feigin, R. D. and Cherry, J. D., Eds., W. D. Saunders, Philadelphia, 1981.
8. **Nassi, L.,** Tentativi di dimostrazione e di coltivazione del virus della quinta malattia, *Riv. Clin. Pediatr.*, 44, 449, 1946.
9. **Werner, G. H., Brachman, P. S., Ketler, A., Scully, J., and Rate, J.,** The fifth disease: an epidemic of acute exanthem in Reading, Pa., *Ann. N.Y. Acad. Sci.*, 67, 388, 1957.

9a. **Smith, J., Grilli, E. and Anderson, M. J.,** unpublished observations.

10. **Balfour, H. H., May, D. B., Rotte, T. C., Phelps, W. R., and Schiff, G. M.,** A study of erythema infectiosum: recovery of rubella virus and echovirus-12, *Paediatrics*, 50, 285, 1972.
11. **Balfour, H. H.,** Fifth disease: full fathom five, *Am. J. Dis. Child.*, 130, 239, 1976.
12. **Laver, B. A., MacCormack, J. N., and Wilfert, C.,** Erythema infectiosum, an elementary school outbreak, *Am. J. Dis. Child.*, 130, 252, 1976.
13. **Anderson, M. J., Lewis, E., Kidd, I. M., Hall, S. M., and Chen, B. J.,** An outbreak of erythema infectiosum associated with human parvovirus infection, *J. Hyg.*, 93, 85, 1984.
14. **Kurtz, J. B. and Anderson, M. J.,** Cross-reactions in rubella and parvovirus specific IgM tests, *Lancet*, 2, 1356, 1985.
15. **Okabe, N., Kobiyashi, S., Tatsuzawa, O., and Mortimer, P. P.,** Detection of antibodies to human parvovirus in erythema infectiosum (fifth disease), *Arch. Dis. Child.*, 59, 1016, 1984.
16. **Shiraishi, H., Wong, D., Purcell, R. H., Shirachi, R., Kumasaka, E., and Numazaki, Y.,** Antibody to human parvovirus in outbreak of erythema infectiosum in Japan, *Lancet*, 1, 982, 1985.
17. **Plummer, F. A., Hammond, G. W., Forward, K., Sekla, L., Thompson, L. M., Jones, S. E., Kidd, I. M., and Anderson, M. J.,** An erythema infectiosum-like illness caused by human parvovirus infection, *N. Engl. J. Med.*, 313, 74, 1985.
18. **Anderson, M. J., Kidd, I. M., and Morgan-Capner, P.,** Human parvovirus and rubella-like illness, *Lancet*, 2, 663, 1985.
19. **Lefrere, J-J., Courouce, A-M., Muller, J-Y., Clark, M., and Soulier, J-P.,** Human parvovirus and purpura, *Lancet*, 2, 730, 1985.
20. **Mortimer, P. P., Cohen, B. J., Rossiter, M. A., Fiarhead, S. M., and Rahman, A. F. M. S.,** Human parvovirus and purpura, *Lancet*, 2, 730, 1985.

20a. **Anderson, M. J.,** unpublished data.

21. **Ager, E. A., Chin, T. D. Y., and Poland, J. P.,** Epidemic erythema infectiosum. *N. Engl. J. Med.*, 275, 1326, 1966.
22. **Reid, D. M., Reid, T. M. S., Brown, T., Rennie, J. A. N., and Eastmond, C. J.,** Human parvovirus-associated arthritis: a clinical and laboratory description, *Lancet*, 2, 422, 1985.
23. **White, D. G., Mortimer, P. P., Blake, D. R., Woolf, A. D., Cohen, B. J., and Bacon, P. A.,** Human parvovirus arthropathy, *Lancet*, 1, 419, 1985.
24. **Lefrere, J. J., Meyer, O., Menkes, C-J., Beaulier, M-J., and Courouce, A-M.,** Human parvovirus and rheumatoid arthritis, *Lancet*, 1, 982, 1985.
25. **Luzzi, G. A., Kurtz, J. B., and Chapel, H. H.,** Human parvovirus arthropathy and rheumatoid factor, *Lancet*, 1, 1218, 1985.
26. **Simpson, R. W., McGuity, L., Simon, L., Smith, C. A., Godzeski, C. W., and Boyd, R. J.,** Association of parvoviruses with rheumatoid arthritis of humans, *Science*, 223, 1425, 1984.
27. Unpublished data quoted in the Communicable Disease Report 84/07 of the Public Health Laboratory Service, U.K.)

Chapter 7

VOLUNTEER STUDIES WITH B19 VIRUS INFECTION

D. A. J. Tyrrell

TABLE OF CONTENTS

I. INTRODUCTION

The story of the discovery of B19 virus resembles in some ways that of some human enteroviruses. Many of these were recognized, it will be recalled, by using the techniques developed to detect polioviruses, namely roller tube cultures of monkey kidney and other cells. By this means cytopathic viruses were often found in the feces of normal children and were called "orphan" viruses — viruses in search of a disease, and so the name ECHO virus was coined. It is worth recalling that period because there was a great deal of emphasis on getting adequate and formal proof that a particular virus was causally related to disease and was not just a harmless virus found by coincidence in a sick patient. In a similar way, virologists of the day took a lot of trouble to prove that many of the "new" respiratory viruses actually caused the diseases in which they were found. Indeed, much thought was given to defining the sort of epidemiological data that should be collected before proof could be regarded as complete. In a number of instances it was possible to fulfill "Koch's postulates" by taking a virus which had been isolated and propagated in the laboratory and inoculating it into volunteers: the object was to show that at least some of those who became infected developed an illness like that seen in infected individuals in the field. This was the purpose of many early studies of the Common Cold Unit.[1] Volunteer experiments have of course also been useful by making it possible to study in detail the exact duration of the incubation period and the association between symptoms and virus shedding etc.

The initial idea behind doing volunteer experiments with B19 virus arose from the early reports that children with sickle cell disease and acute erythroid aplasia showed evidence of infection in that they often gave a history of respiratory symptoms a week or so before they were admitted.[2,3] It was therefore thought that B19 virus might be a previously unrecognized cause of respiratory disease and that this could be proved by giving the virus to normal subjects in isolation. By the time the studies took place it was also clear that a closely related or identical virus was associated with "fifth disease", erythema infectiosum, and that some adults with the disease had joint involvement: although the virus had not been detected in such patients almost all had specific IgM antibodies.[4] Thus the questions changed as the plan of the studies developed but at the time they began a number of answers were needed:

1. Did the virus found in the blood infect man if given by the intranasal route?
2. If it did, did it cause viremia and was it shed externally?
3. If infection occurred was there an effect on the bone marrow, a rash, or joint involvement?
4. Was there specific immunity and what was the relationship between virus, immune response, and any disease?
5. Did the virus cause respiratory symptoms?
6. Did the *same* virus cause symptomless viremia, affect the marrow, and cause a rash?

II. THE CONDUCT AND RESULTS OF THE FIRST TRIAL

After a review by the ethical committee, it was agreed that the proposed experiments were likely to produce scientifically useful results and carried negligible risk, and so they were undertaken. Individuals who had already been to the Unit as subjects for experiments with respiratory viruses were specially approached. They had to agree to stay at the Common Cold Unit for 3 weeks and to accept the chance of having a rash, etc. Any who gave a history of blood or joint disease in themselves or in close relatives were not accepted, nor were those who gave a history of hay fever or other allergies. The double-blind for clinical observations was maintained by including some volunteers with antibodies who were not expected to become infected — volunteers were told this was the case but neither they nor the clinical observer was told who was seropositive and who was not.

However, before the main study in isolated volunteers began two of the scientific team, one with serum antibodies and another with only a trace were inoculated.[5] The second became infected and developed transient fever, malaise, and blood changes. Once these preliminary results were obtained the main study began. Small groups were inoculated in succession but the results of the whole group are presented together here, except that part of the data from the two initial volunteers had to be excluded since their clinical observations were purely subjective and their hematological results were very variable as they were not living in the stable and protected environment of the Unit.

In the Unit, the volunteers remained in isolation for the first few days and were found healthy by clinical examination and by routine biochemical and hematological tests. They were inoculated by giving nasal drops containing infected serum diluted in buffered salt solution. The serum had been found by screening routine blood donations and had been shown to be free of bacteria, fungi, and other viruses by inoculation into standard laboratory test systems. Clinical observations continued daily, concentrating particularly on the respiratory tract, abdomen, skin, and joints. Nasal washing, urine, feces, and blood were collected at regular intervals for virological study, using methods referred to in Chapter 4.

A. Virological Events

The results of the studies are summarized first in Table 1 which shows that five subjects with low (PV1) or undetectable antibodies (V3-6) became infected whereas three with circulating antibody (PV2, V1-2) did not. Serum diluted 100-fold was infectious for V5 and V6, whereas the one seronegative individual given a 10^{-4} dilution was not infected as judged by either virus detection or immune response. All the infected volunteers showed a viremia followed by an IgM and IgG response in circulating antibody. One volunteer (PV1) who had a trace of IgG antibody at the time of inoculation developed less IgM and an early and substantial IgG response and this suggested he was making a secondary response — rather than the primary response of other volunteers. The uninfected volunteers showed only very mild and transient clinical symptoms which were not regarded as significant whereas most infected volunteers had definite signs and symptoms, and this combined with the fact that they were isolated from intercurrent infection is strong evidence that what was observed was due to the virus infection.

The magnitude and time course of these changes are shown in the figures. It should be noted that the viremia occurred at a very similar time in all cases and reached high titers — up to 10^{-5} dilution was positive on dot hybridization assay (Figure 1). Virus was also detected in the upper respiratory tract at the time of viremia. The IgM response began just as the viremia was declining and may indeed be the reason for the decline (Figure 2). Certainly IgM immune complexes were detected at that time, although usual tests for immune complexes, such as Clq binding were negative. The serum of all infected volunteers contained specific IgG antibody by the time the late serum was collected, 2 to 3 weeks after leaving the Unit and about 6 weeks after inoculation (Figure 3). However, some volunteers developed such an antibody while they were still at the Unit, but still many days after IgM had been detected.

B. Clinical Illness

The illness pattern seen was biphasic. The viremia could be asymptomatic but was also associated with malaise, myalgia, or transient fever. This relationship can be seen from Figure 4 which summarizes the observations made on one volunteer (V5). The second phase of illness included a rash and joint involvement. The rash occurred particularly on the trunk and limbs, and was maculopapular in appearance and lasted 2 or 3 days. It was not distinctive but compatible with what has been seen in adults in outbreaks of fifth disease. The joint involvement consisted of mild pain and stiffness, particularly of the joints of the hands and

Table 1
INOCULATION DATA AND EVIDENCE OF HPV INFECTION IN VOLUNTEERS

Volunteers	Sex	Virus inoculum (dilution of serum)	Preinoculation anti-HPV (a.u.)[a]	Illness	Viremia	Postinoculation IgM anti-HPV	Postinoculation IgG anti-HPV	Postinoculation Virus shedding
PV1	M	10^{-1}	0.5	Fever	+	+	+	–
PV2	M	10^{-1}	2.0	Nil	–	–	–	–
V1	F	10^{-1}	4.0	Nil	–	–	–	–
V2	M	10^{-1}	2.6	Nil	–	–	–	–
V3	F	10^{-1}	<0.3	Mild "flu" R & J[b]	+	+	+	+
V4	M	10^{-1}	<0.3	Mild "flu"	+	+	+	–
V5	F	10^{-2}	<0.3	Mild fever R & J	+	+	+	+
V6	F	10^{-2}	<0.3	Malaise R & J	+	+	+	+
V7	F	10^{-4}	<0.3	Nil	–	–	–	–

[a] Arbitrary units.
[b] Second phase of illness with maculopapular rash, pain, and stiffness of joints. Note that all those with this syndrome were women.

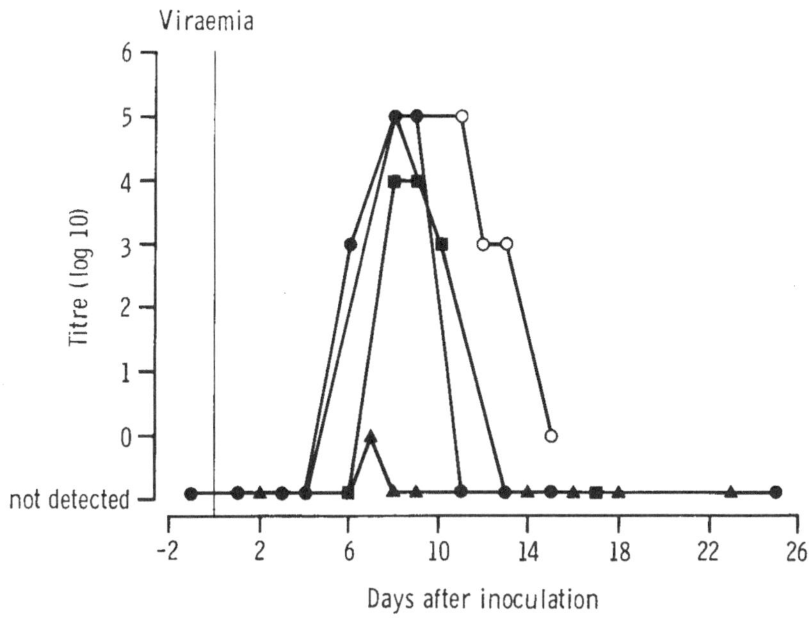

FIGURE 1. Titer of HPV DNA detected by dot-blot hybridization following intranasal inoculation of seronegative volunteers (V3-●, V4-○, V5-■, V6-□) and a volunteer (PV1) with trace amounts of antibody (▲——▲). (From Anderson, M. J., Higgins, P. G., Davis, L. R., William, J. S., Jones, S. E., Kidd, I. M., Pattison, J. R., and Tyrrell, D. A. J., *J. Infect. Dis.*, 152, 257, 1985. With permission.)

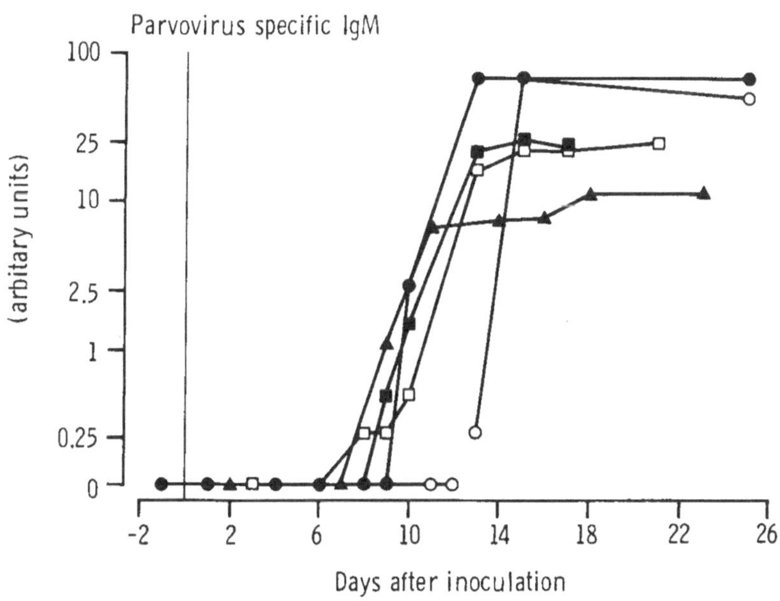

FIGURE 2. Development of HPV-specific IgM following inoculation of volunteers (symbols as above). (From Anderson, M. J., Higgins, P. G., Davis, L. R., William, J. S., Jones, S. E., Kidd., I. M., Pattison, J. R., and Tyrrell, D. A. J., *J. Infect. Dis.*, 152, 257, 1985. With permission.)

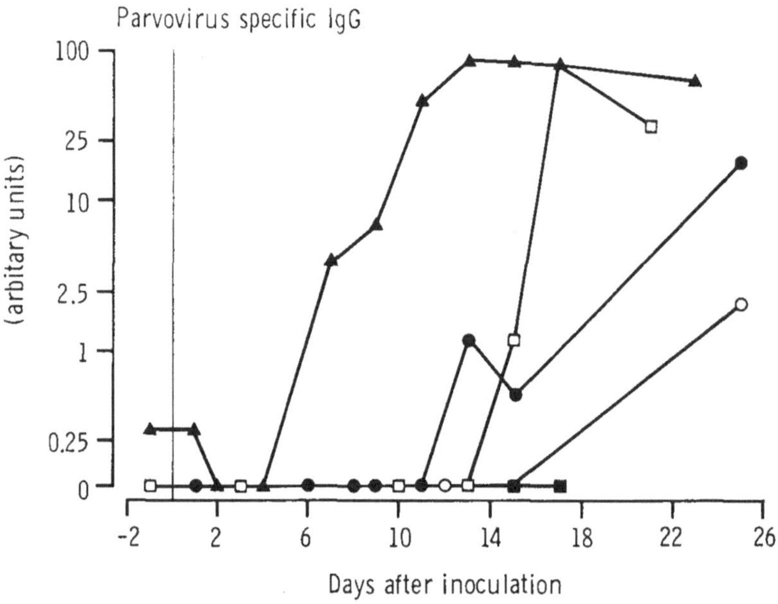

A

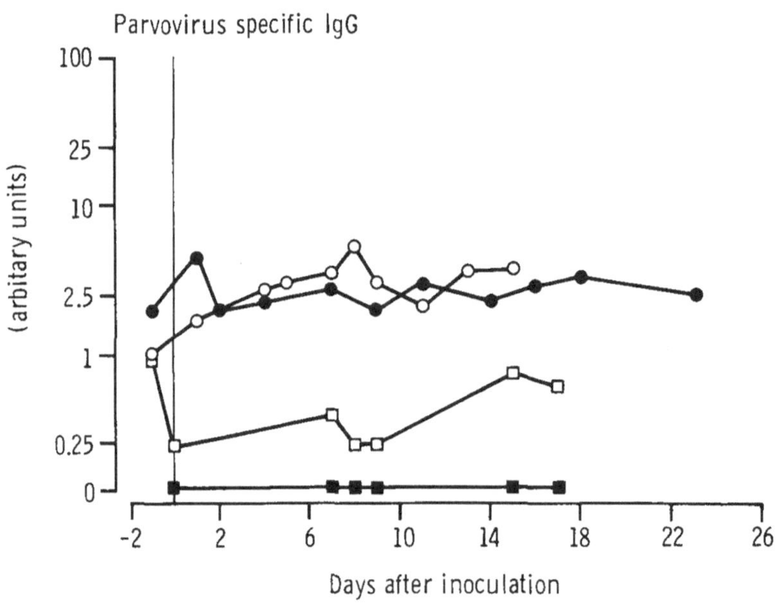

B

FIGURE 3. (A) Development of HPV-specific IgG following intranasal inoculation of seronegative volunteers and a volunteer with trace amounts of antibody (symbols as in Figure 1); (B) Lack of response in uninfected volunteers (PV2-●, V1-○, V2-□, V7-■). (From Anderson, M. J., Higgins, P. G., Davis, L. R., William, J. S., Jones, S. E., Kidd, I. M., Pattison, J. R., and Tyrrell, D. A. J., *J. Infect. Dis.*, 152, 257, 1985. With permission.)

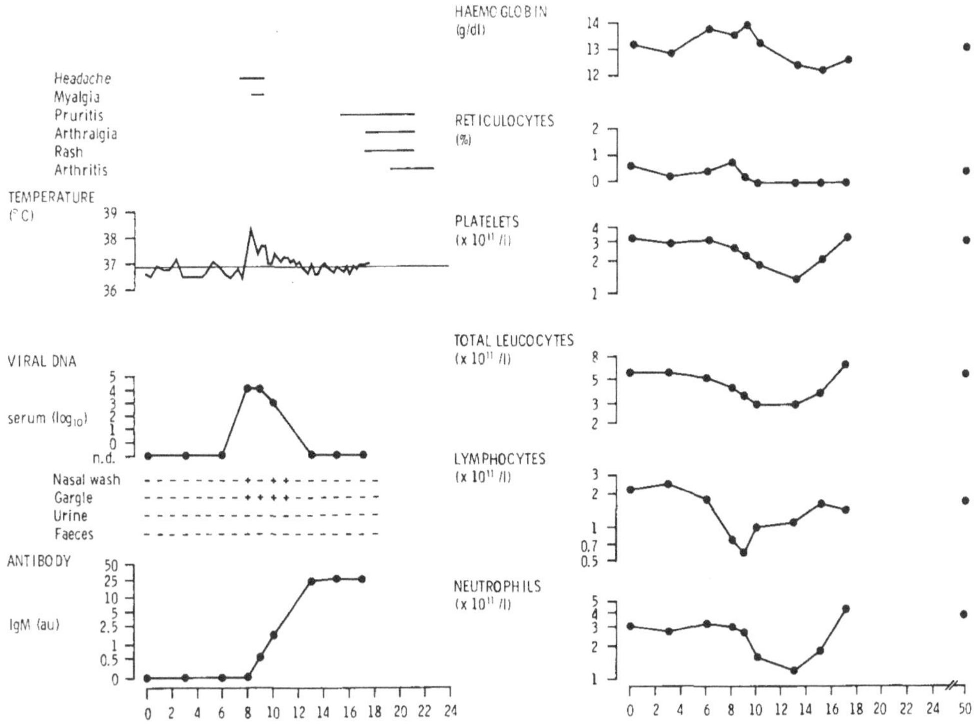

FIGURE 4. Features of primary parvovirus infection in a seronegative volunteer (V5) inoculated intranasally on day 0. The biphasic disease is shown (top left) with the evidence for virus infection and shedding and the immune response. On the right is the evidence for interruption of erythropoiesis and changes in white cell count. (From Anderson, M. J., Higgins, P. G., Davis, L. R., William, J. S., Jones, S. E., Kidd, I. M., Pattison, J. R., and Tyrrell, D. A. J., *J. Infect. Dis.*, 152, 257, 1985. With permission.)

feet. There was some swelling of the surrounding tissue, particularly noticeable in the fingers, and this lasted rather longer than the rash. The course of the later disease was not observed closely as the volunteers were released from the Unit at or soon after the time it started — they had agreed to stay only 3 weeks and the illness seemed mild and to be resolving spontaneously. The data provide useful estimates of the incubation period of both syndromes.

C. Hematological Changes

Clear-cut abnormalities were also observed in the hematological examination of the isolated patients. Reticulocytes disappeared from the peripheral blood of all four infected volunteers and were absent for over a week and during this time the hemoglobin concentration fell by 1.2 to 1.8% per day (Figure 5). These volunteers also showed a neutropenia during the second week after which cell counts returned to slightly above those obtained before inoculation. There was also a decline in lymphocytes in infected volunteers. In these four volunteers there was also a decline in platelet counts although they remained within the accepted normal range and then returned to preinoculation values after about 1 week. No such changes occurred in the uninfected volunteers. Neither infected nor uninfected volunteers showed any changes in blood bicarbonate, calcium, sodium, potassium, inorganic phosphate, urea, creatinine, urate, total protein, albumin, total bilirubin, alkaline phosphatase, aspartate transaminase, or gammaglutamyl transferase.

III. COMMENT

This set of results helps to answer the questions mentioned earlier. In the first place it

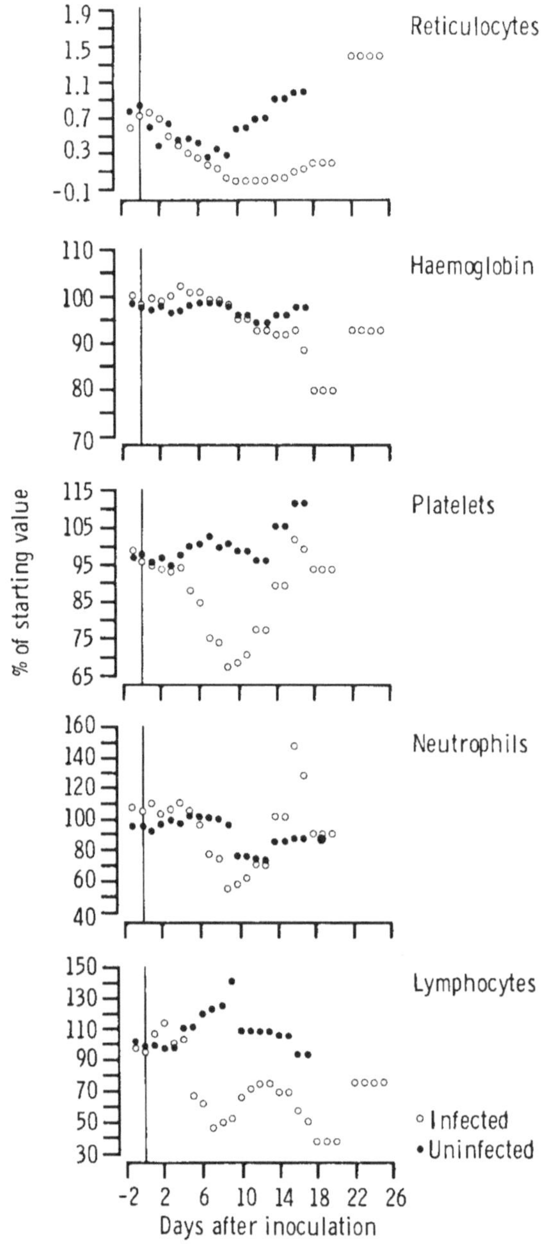

FIGURE 5. The top panel shows reticulocyte counts ($\times 10^{-6}$) in (○) infected and (●) uninfected volunteers. Below it is shown hemoglobin concentration, in infected (○) and (●) uninfected volunteers, expressed as a percentage of the initial values obtained prior to intranasal inoculation on day 0. The lower three panels are rolling mean values of platelets and white blood cells in infected (○) and uninfected (●) volunteers. Each point represents a value computed by calculating the mean value for all volunteers in a group, of each data point within a 4-day period. The data points used in this calculation were the cell counts expressed as a percentage of the initial values prior to inoculation for that volunteer. (From Anderson, M. J., Higgins, P. G., Davis, L. R., William, J. S., Jones, S. E., Kidd, I. M., Pattison, J. R., and Tyrrell, D. A. J., *J. Infect. Dis.*, 152, 257, 1985. With permission.)

proved that virus particles found in the blood are infectious if given intranasally. Of course such drops are in part swallowed and we do not know where infection begins, but we can suggest that the pharynx is the susceptible site since it is there that virus appears later in infection. Viremia occurred, followed by an immune response just as in the cases of aplastic crisis complicating sickle cell disease. It is therefore plausible that infection of such children enters by the respiratory route. However, the volunteers showed no respiratory symptoms and it may be that the reports of such symptoms in sickle cell disease children represent coincidence, or alternatively that infection is symptomatic in them but not in normal adults. Since the virus was not found in feces or urine, it is likely to be transmitted via respiratory secretions.

The two distinct phases of the illness clarified the apparent conflicts in the clinical accounts. Presumably the rash phase has been absent, or not noticed in the children with erythroid aplasia and sickle cell disease. Normal subjects can well tolerate lack of erythropoesis for a week or so and therefore in such individuals erythroid aplasia in the week before erythema infectiosum will be asymptomatic. Since the inoculum came from a normal blood donor we can also conclude that one and the same virus can cause subclinical infections and both clinical syndromes and that there is no evidence for three sorts of virus each with a different pathogenic potential.

There are still immunological questions to be answered. It is clear that circulating antibody is a good index of resistance to infection and apparently clears the viremia, but other immune systems may be operative also. It is of interest that IFNα was not found in the circulation but cellular immunity might be active. We believe that the second phase of illness is immune-mediated, but we have no specific information. Perhaps it is a response to immune complexes lodged in the skin, but they have not been shown to be present and the inflammatory response has not been seen microscopically.

The pathogenesis of the blood changes is also quite unclear. Is there marrow aplasia? Is the change due to virus present in marrow cells or to an effect on the control system, perhaps erythropoietin? Are there different changes in the different lymphocyte subsets? Are the polymorphs marginated on blood vessels?

In view of these unanswered questions it was desirable to make further and more detailed observations on other volunteers infected in a similar way.

IV. THE SECOND TRIAL

On the same basis as mentioned earlier three male volunteers were recruited for a study of the changes in the erythroid and myeloid precursors present in the marrow and peripheral blood after intranasal infection with the same pool of virus as was used previously.[6]

A mild disease was produced as before and there was viremia which reached a peak 9 days after infection and typical changes in the peripheral blood counts were seen. Volunteers agreed to iliac crest marrow aspirates on three occasions, before, during, and after viremia and these showed that at 10 days there was a gross reduction in erythroid precursors which was restored by 10 weeks after infection. Following the erythroid aplasia and decline of peripheral blood hemoglobin there was an increase in circulating erythropoietin by immunoassay. Peripheral blood cells were cultured and there was a clear-cut decline in both BFU-E and CFU-GM at the time of viremia and for a few days after (Figure 6). This then recovered and there was a considerable "overshoot" in the counts when recovery commenced. Marrow cells were also cultured but the changes were less marked and were not consistent, but in the individual who had the longest period of reduced counts in the peripheral blood there was also a reduction in the progenitor cells recovered from the marrow.

Lymphocytes were also examined in some detail and there were no changes in the distribution of lymphocyte subsets although the total counts fluctuated as already described.

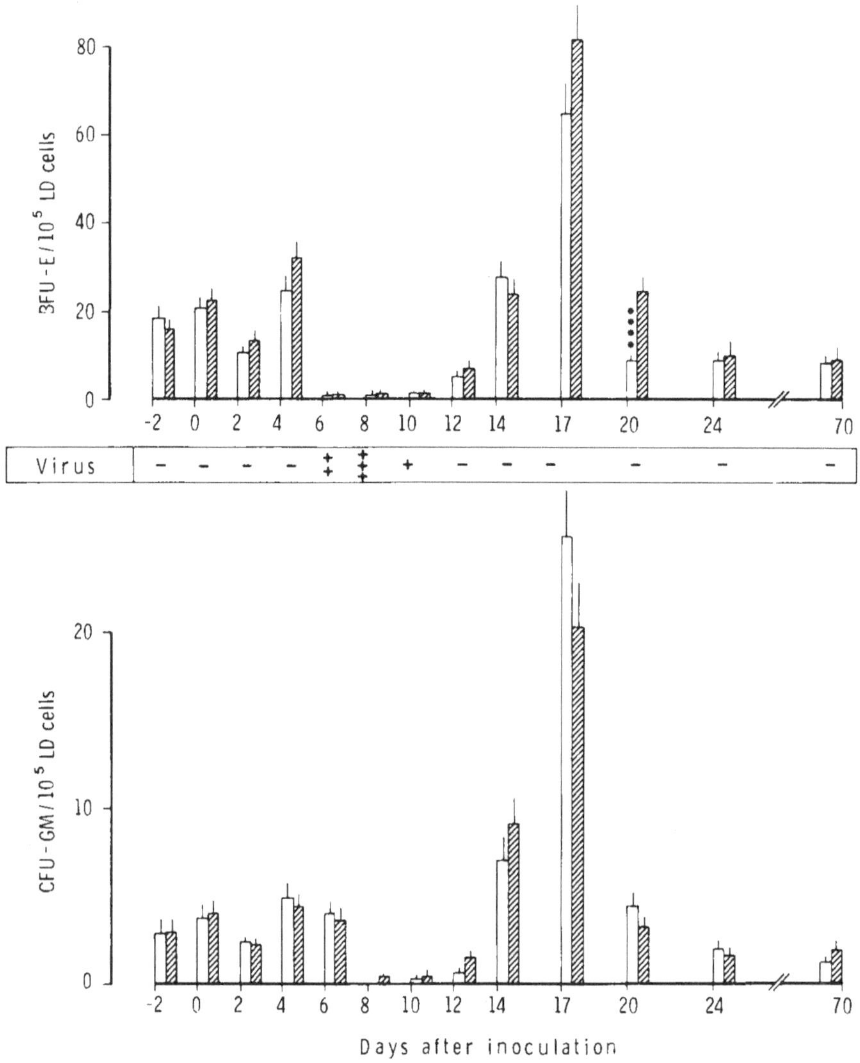

FIGURE 6. Growth of BFU-E and CFU-GM from the peripheral blood of a volunteer inoculated with B19 virus on day 0. The hatched columns indicate samples cultured with 10% autologous serum and the open columns without. (Reproduced from Potter, C. G., Potter, A. C., Hatton, C. S. R., Chapel, H. M., Anderson, M. J., Pattison, J. R., Tyrell, D. A. J., Higgins, P. G., Willman, J. S., Parry, H. F., and Cotes, P. M., *J. Clin. Invest.*, 79, 1486, 1987 by copyright permission of the American Society for Clinical Investigation.)

V. CONCLUSION

These studies thus answered some of the questions asked earlier, but more work is needed to discover the pathogenesis of the rash and joint disease, and also to work out where the virus found in the blood is replicated and how it produces such a profound and reversible arrest of red cell production.

REFERENCES

1. **Tyrrell, D. A. J.,** *Common Colds and Related Diseases*, Arnolds, London, 1965.
2. **Pattison, J. R., Jones, S. E., Hodgson, J., Pavis, L. R., White, J. M., Stroud, C. E., and Mustaza, L.,** Parvovirus infections and hypoplastic crisis in sickle cell anaemia, *Lancet*, 1, 664, 1981.
3. **Serjeant, G. R., Mason, K., Topley, J. M., Serjeant, B. E., Pattison, J. R., Jones, S. E., and Mohamed, R.,** Outbreak of aplastic crisis in sickle cell anaemia associated with parvovirus-like agent, *Lancet*, 2, 595, 1981.
4. **Anderson, M. J., Lewis, E., Kidd, I. M., Hall, S. M., and Cohen, B. J.,** An outbreak of erythema infectiosum associated with human parvovirus infection, *J. Hyg.*, 92, 85, 1984.
5. **Anderson, M. J., Higgins, P. G., Davis, L. R., William, J. S., Jones, S. E., Kidd, I. M., Pattison, J. R., and Tyrrell, D. A. J.,** Experimental parvoviral infection in humans, *J. Infect. Dis.*, 152, 257, 1985.
6. **Potter, C. G., Potter, A. C., Hatton, C. S. R., Chapel, H. M., Anderson, M. J., Pattison, J. R., Tyrrell, D. A. J., Higgins, P. G., Willman, J. S., Parry, H. F., and Cotes, P. M.,** Variation of erythroid and myeloid precursors in the marrow and peripheral blood of volunteer subjects infected with human parvovirus (B19), *J. Clin. Invest.*, 79, 1486, 1987.

Chapter 8

STUDIES OF THE B19 VIRUS IN BONE MARROW CELL CULTURE

Neal Young and Keiya Ozawa

TABLE OF CONTENTS

The availability of well-characterized tissue culture systems for bone marrow cells has proved fortunate for B19 virus studies. To date, culture of human bone marrow has provided the only in vitro technique for either examination of the cellular and molecular events associated with B19 infection for propagation of the virus. The study of the interaction of this specific virus with hematopoietic progenitor cells has also proved to be illuminating to hematologists interested in specific mechanisms of bone marrow failure. This chapter will describe the hematopoietic culture systems employed, the outcomes of studies of B19 virus in both clonal and suspension cultures, and the implications these results have for the mechanism of B19 virus infection and hematopoietic failure in general.

I. HEMATOPOIESIS IN VIVO AND IN VITRO[1,2]

The bone marrow, in contrast to most visceral organs, is a continuously proliferating tissue. The cellular elements of the peripheral blood — erythrocytes, leukocytes, and platelets — have limited life spans and are replenished by the differentiation and proliferation of more immature cells in the marrow.

Three types of immature cells can be distinguished (Figure 1). Immature precursors of erythrocytes, granulocytes, and platelets are morphologically recognizable in stained preparations of bone marrow as, for example, erythroblasts, myelocytes, and megakaryocytes. The mitotic activity of precursor cells can be grossly demonstrated by uptake of labeled thymidine in bone marrow suspension cultures. Precursors are in turn derived from cells called committed progenitors; these cells cannot be recognized by their appearance but have been defined by their ability to form colonies when assayed in vitro. Committed progenitors are derived from hematopoietic stem cells; the hematopoietic stem cell is the ultimate origin of circulating cells. Hematopoietic stem cells have been demonstrated in animal models and in culture systems, as well as inferred from the clonal origin of many hematologic diseases.

The limited number of stem cells present at birth and their continued loss to terminal differentiation requires proliferation to occur within the stem cell compartment; the continued generation of stem cells is termed self-renewal. Within the stem cell pool there is considerable heterogeneity, as measured by self-renewal, differentiation, and proliferative capacities. Some stem cells become committed to differentiate (and proliferate) along selected, lineage-specific pathways to produce erythrocytes, leukocytes, and platelets. The balance between stem cell self-renewal and differentiation and the ordered specialization and expansion of cells in differentiation pathways are both highly regulated. Proliferative and differentiative events along committed pathways, and probably also within the stem cell population, are regulated by factors collectively termed colony stimulating activities. Although these factors have been studied as soluble additives to colony stimulating assays, they may in fact function in animals at the cell surface and account for the required cell-cell interactions of hematopoiesis.

The existence of stem cells was inferred from experimental observations on sublethally irradiated mice that had received suspensions of spleen or bone marrow cells. Not only was hematopoiesis restored in these animals, but large colonies formed in the spleen; these colonies could be demonstrated to be single-cell in origin. Within colonies were morphologically recognizable precursor cells for erythrocytes, granulocytes, platelets, and lymphocytes. Also present within spleen colonies were cells capable of giving rise to new colonies in secondary transplantations. The cell detected in the spleen colony assay, termed CFU-S (colony forming unit-spleen) therefore had the expected properties of a hematopoietic stem cell: multipotency, self-renewal, and high proliferative capacity.

The CFU-S assay has been criticized both theoretically and practically. Inability to serially passage cells from CFU-S-derived colonies beyond a few transplants, in contrast to the ability of small numbers of marrow or spleen cells to support hematopoiesis in intact animals, has suggested the presence of a hematopoietic stem cell even more primitive than the CFU-

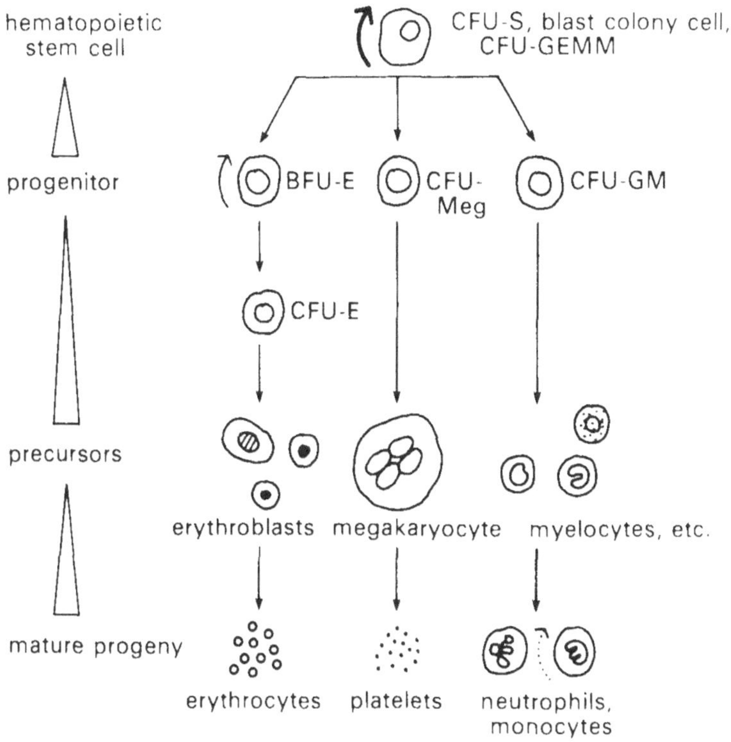

FIGURE 1. A schema of hematopoiesis. Representative cells from the compartments of morphologically recognizable precursors are shown. Amplification of cell number by mitosis occurs at every stage, except in the later stages of terminal differentiation (for example, after the myelocyte stage in granulocyte differentiation; after the polychromatophilic normoblast stage in erythropoiesis). As a simplification, the generation of eosinophils, mast cells, and lymphocytes from the same common stem cell is not shown. Self-renewal capacity is denoted by the curved arrows. For further explanation and definition of abbreviations see text.

S. CFU-S determination also rests on several modeling assumptions, as for example, those related to seeding efficiency. CFU-S assays are cumbersome to perform, and of course they cannot be applied to human cells. The measurement of progenitor cells committed to a pathway of differentiation in vitro overcame these practical problems. While colonies derived from hematopoietic stem cells (CFU-GEMM, "blast" colonies; see below) have been detected with difficulty using in vitro clonal assays, in most situations the good correlation of stem cell with progenitor number and function permits the use of standard, more easily established committed progenitor systems.

A. Clonal (Progenitor) Assays

Immobilization of bone marrow cultures by the addition of a semisolid additive, usually methylcellulose, agar, or clotted plasma, allowed definition of hematopoietic progenitor cells by their ability to form colonies. These cultures are purposely performed at low plating concentrations of cells, and the resulting hematopoietic colonies have been demonstrated as single cell in origin by cytogenetic or enzyme analysis. Colony counting is quantitative, but the necessarily low plating concentration used in clonal assays requires the creation of an artificial environment for optimal growth of progenitor cells. Dilution of cells disturbs the intimate relationship among different types of bone marrow cells, and optimal colony formation has been found empirically to require addition of constituents that promote cell viability and proliferation, most usually fetal calf serum, bovine serum albumin, and medium conditioned by lectin-stimulated leukocytes or placenta tissue.

Differentiation and proliferation are linked in most clonal assays. Colonies committed to erythroid (BFU-E, CFU-E), myeloid (granulocyte-macrophage; CFU-GM), or megakaryocyte differentiation can generally be distinguished microscopically. Differences in the kinetics of colony formation and dependence on specific growth factors have permitted distinction between erythroid colony forming cells of greater or lesser degrees of maturity. First, there is the primitive burst forming cell (BFU-E) which generates large colonies, composed of thousands of normoblasts, after weeks in culture in the presence of high concentrations of erythropoietin and burst promoting activity. Secondly, there is the more mature erythroid colony forming cell (CFU-E), which forms colonies of 50 to 100 cells 5 to 7 days after plating with minimal growth factor requirements; this cell is much more sensitive to the addition of erythropoietin than the BFU-E.

B. Long-Term (Dexter) Cultures

Cell-cell contacts are purposely promoted in long-term (Dexter) cultures,[3] which rely on the formation of an adherent cell layer at the bottom of a flask to support hematopoiesis, measured by the release of progenitors (separately assayed in clonal cultures) into the nonadherent cell layer. The physical relationship of cells in the adherent cell layer probably mimics that in the bone marrow stroma. Long-term cultures, particularly murine, have been especially useful in the study of hematopoietic stem cell self-renewal, although they generate data more qualitative than quantitative. This is especially true of long-term cultures of human bone marrow, which have not been adapted to reach "steady-state" conditions and show declining total cell and progenitor cell numbers from the time of initiation. The suspension cultures utilized for studies with B19 virus are not long-term cultures because the growth conditions and time of incubation are inadequate for creation of an adherent cell layer.

II. EFFECTS OF B19 VIRUS ON CLONAL PROGENITOR CELL ASSAYS

Although B19 virus had been associated by seroepidemiological studies with transient aplastic crises, the mechanism by which the virus affected red blood cell production was elucidated with in vitro studies. Virus infections so often alter peripheral blood counts, reflected in the common occurrence of neutropenia and the occasional development of thrombocytopenia, that "nonspecific" effects have usually been invoked to explain bone marrow depression; in fact, bone marrow depression may result from the generation of interferons in the course of viral infections. Study of the action of B19 virus on clonal hematopoietic assays showed instead that the virus had a highly selective and specifically cytotoxic effect on a single class of progenitor cells.

Addition of sera containing B19 virus to clonal assays of normal bone marrow dramatically inhibited colony formation by erythroid progenitor cells.[4,5] Myeloid colony formation was not affected at dilutions of sera that consistently blocked the appearance of CFU-E-derived colonies. These sera were extraordinarily potent inhibitors of erythropoiesis; the amounts of virus added were necessarily small because of the limited quantities of sera available. In contrast to experiments with antibody inhibitors of hematopoiesis, in which sera usually are added undiluted to 10% final volume of culture mixes, dilutions of virus-containing sera to 1 in 10^4 significantly inhibited erythropoiesis. The specificity of the B19 virus for erythroid colony forming cells and the potency of sera containing the parvovirus as an inhibitor of erythropoiesis in vitro are illustrated in Figure 2.

That parvovirus in the sera inhibited erythropoiesis was suggested by the ability of convalescent phase sera (containing antibody to B19 virus) to partially or completely abrogate the inhibition by B19 of erythroid colony formation by CFU-E.[4] In contrast, antisera to

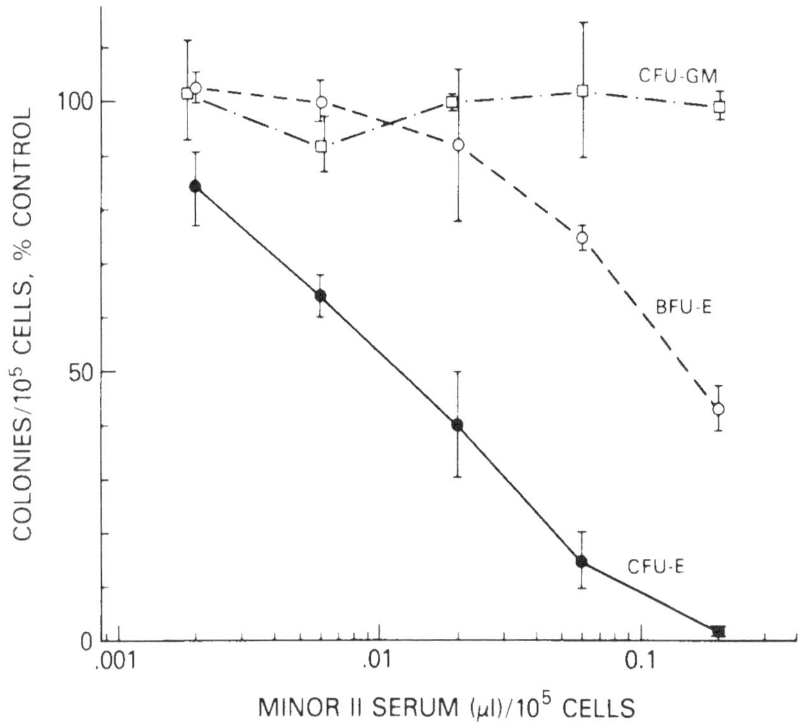

FIGURE 2. Inhibition of colony formation by committed progenitor cells in normal human bone marrow cultures. For these experiments, 2×10^7 mononuclear cells were incubated with varying quantities of B19-containing serum (Minor II) for 2 hr prior to culture. For comparison with previously published data,[4,6] 0.05 μℓ antigen = 10^{-2} dilution of sera (based on the assumption that multiplicity of infection is related to amount of virus/cell number and unaffected by concentration of cells or virus in this narrow range).

human IgG or human alpha interferon did not affect colony inhibition. Most important, fractionation of sera by sucrose density gradient centrifugation localized inhibitory activity to the fraction containing virus, while the IgG and IgM fractions were not inhibitory.[6]

The formation of hematopoietic colonies is a complex process, and proliferation of progenitor cells can be shown to be dependent upon appropriate growth factors or the cells that produce growth factors. Particularly as colony inhibition by a virus was unprecedented in clonal assays, B19 infection of an accessory cell, required for erythroid colony development, was an alternative mechanism of virus action. However, the effect of B19 virus directly on erythroid progenitors could be demonstrated in several ways. First, the effect of virus on a relatively pure population of late erythroid progenitors was assayed. CFU-E are represented in normal bone marrow at a low frequency (0.05 to 0.4% of mononuclear cells). Progenitors with equivalent growth properties to CFU-E can be isolated by visual identification and manual removal from primary culture of immature bursts (BFU-E-derived colonies at 7 to 8 days); in fresh semisolid medium, these cells form only secondary erythroid colonies (greater than 10% plating efficiency). Incubation of sera containing B19 virus with plucked bursts prior to reculture completely eliminated secondary erythroid colony formation, and examination of culture dishes during the first 2 days after replating showed failure of colony formation to proceed beyond the two to four cell stage.[6]

Isolation of erythroid progenitors, infection with virus, and processing 48 hr later allowed morphologic demonstration of B19 virus.[7] By immunofluorescence with a monoclonal antibody directed to B19 virus 8, approximately 10 to 20% of cells in infected cultures showed specific cytoplasmic fluorescence in a particulate pattern; nuclei were less commonly positive

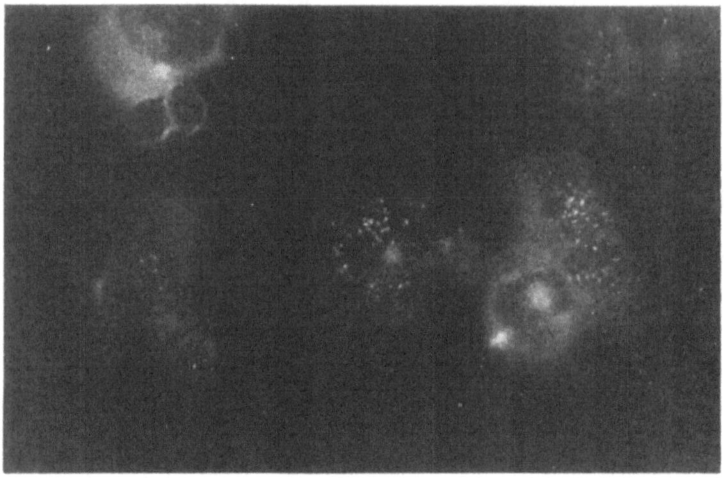

FIGURE 3. B19 virus-specific immunofluorescence. Normal bone marrow cells were cultured in the presence of erythropoietin and PHA-leukocyte conditioned medium to promote erythropoiesis. Immature bursts were identified morphologically in primary culture and removed by suction with a finely tapered Pasteur pipette. (These cells represented an enriched (10 to 30%) population of CFU-E.) Cells were exposed to serum containing B19 virus and then replated in fresh medium and methylcellulose. Cells were harvested from secondary plates 48 hr later, concentrated by cytocentrifugation, fixed in acetone, and stained with a mouse monoclonal antibody to B19 capsid proteins.[7] (Magnification ×840) (From Young, N. S., Harrison, M., Moore, J. G., Mortimer, P. P., and Humphries, R. K., Direct demonstration of the human parvovirus in erythroid progenitor cells infected in vitro, *J. Clin. Invest.*, 74, 2024, 1984. With permission.)

(Figure 3). This pattern was consistent with semipermissive growth of virus in its target cell. On electron microscopy, infected cultures contained many abnormal cells, characterized by pseudopod formation, vacuolization, abnormal mitochondria, and marginated nuclear chromatin (Figure 4). Within the margins of chromatin were lacunae containing arrays of crystalline particles 18 nm in diameter, highly suggestive of parvovirus structures (Figure 5). Because their shape and size were similar to ribosomes, parvovirus could not be definitively detected in cytoplasm.

III. PROPAGATION OF B19 VIRUS IN BONE MARROW CULTURES

Clonal assay of human bone marrow clearly showed that the B19 virus infected erythroid progenitor cells, but this culture system was inherently limited by the small number of cells purposely employed to permit accurate colony enumeration. For purposes of propagation of virus — to allow production of virus and to determine the molecular events during virus replication — we and others tested a wide variety of cell lines, all with negative results.

However, B19 virus was shown to propagate in suspension cultures of bone marrow.[9] The apparent specifity of the virus for erythroid progenitor cells led to the choice of erythroid bone marrows, obtained from patients with sickle cell disease. While normal bone marrow contains about 30% identifiable erythroid cells, sickle cell disease bone marrow has an erythroid content of 50 to 70%. Addition of serum containing B19 virus to cultures of mononuclear cells from the bone marrow of patients did not significantly lower the total cell number at the end of 1 week of culture, but the number of erythroid cells in infected cultures was generally less than 10% of erythroid cells in control cultures at this time.

Detection of virus and of viral replication was facilitated by the cloning of the B19 virus

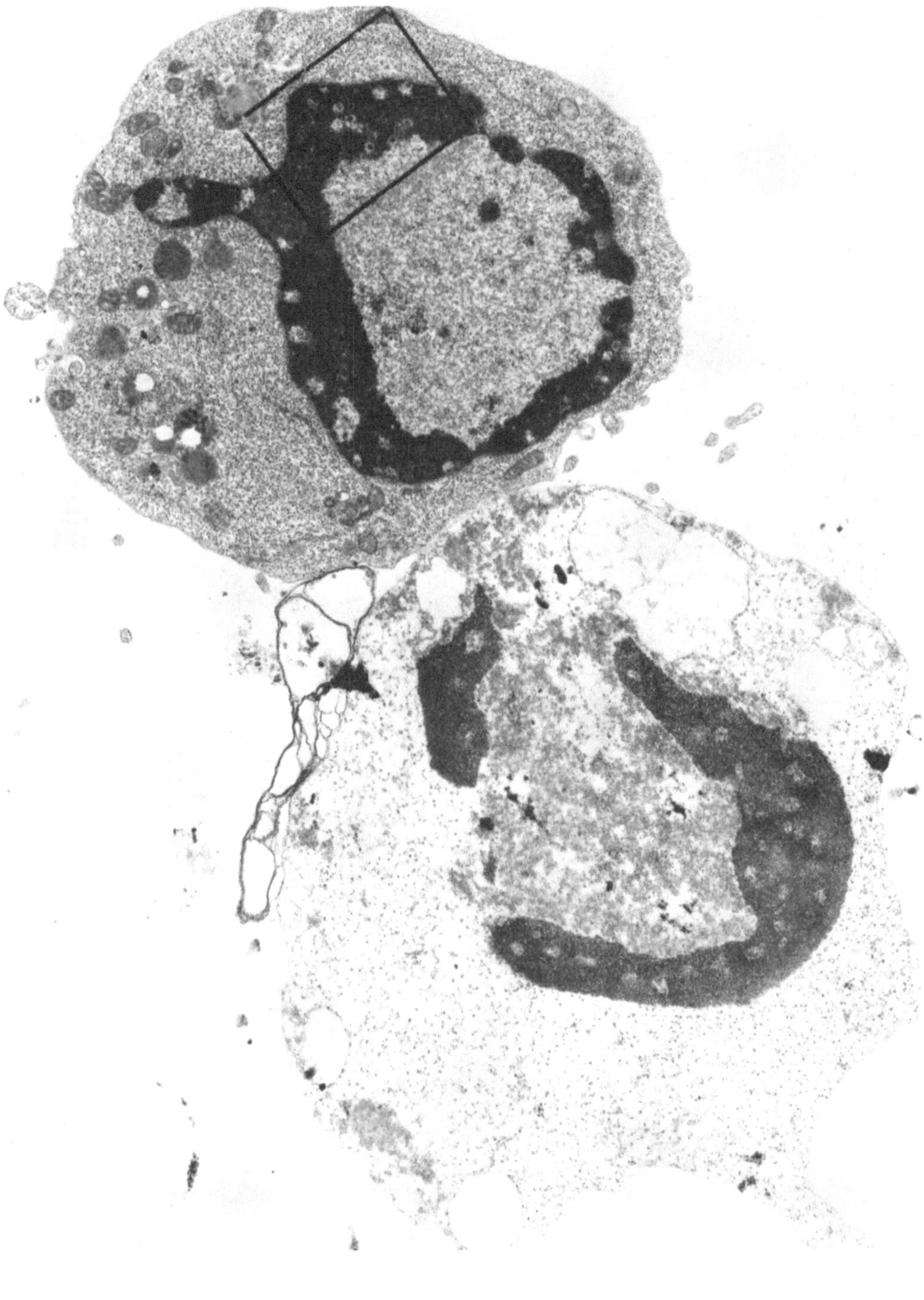

FIGURE 4. Morphologic appearance of B19 infected cells. Cells were prepared as in the legend to Figure 3 and examined by electron microscopy.[7] (Magnification × 12,000) The enclosed area is shown at higher magnification in Figure 5. (From Young, N. S., Harrison, M., Moore, J. G., Mortimer, P. P., and Humphries, R. K., Direct demonstration of the human parvovirus in erythroid progenitor cells infected in vitro, *J. Clin. Invest.*, 74, 2024, 1984. With permission.)

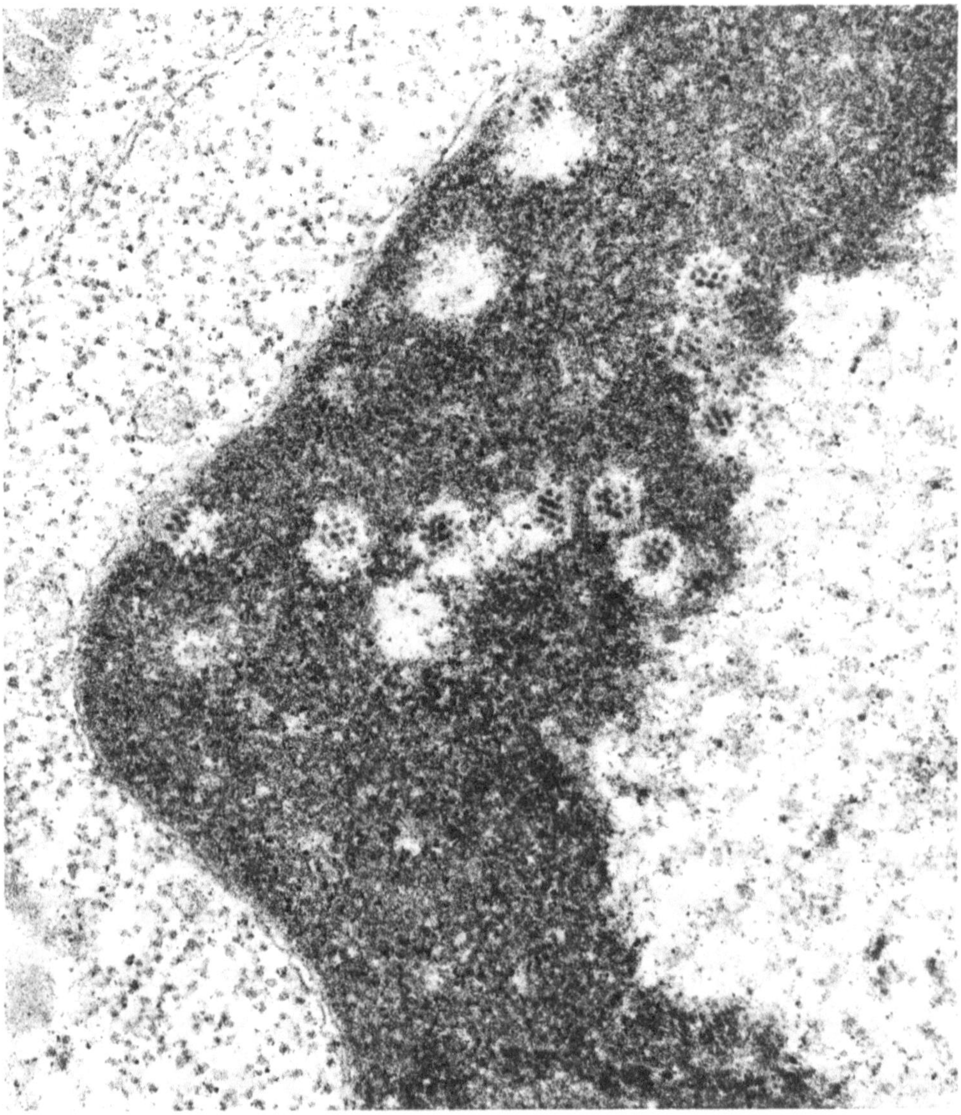

FIGURE 5. B19 virus in infected progenitor cells. (Original magnification ×99,000) (From Young, N. S., Harrison, M., Moore, J. G., Mortimer, P. P., and Humphries, R. K., Direct demonstration of the human parvovirus in erythroid progenitor cells infected in vitro, *J. Clin. Invest.*, 74, 2024, 1984. With permission.)

genome[10] and the availability of a full length recombinant construct, pYT103 (a generous gift of P. Tattersall and S. Cotmore).

For these experiments, DNA was extracted from whole cells, cytosol, and nuclei from bone marrow cultures infected 40 hr previously with B19 virus. To detect B19 virus specific sequences, the DNA was blotted onto nitrocellulose paper and probed with a nick translated, radioactive pYT 103 probe. Strong signals indicative of the presence of B19 virus-like sequences were detected in the whole cell and also in the nuclear preparations. The presence of more B19 virus DNA in the nuclei than in the cytosol suggested active replication of the virus.

The replication strategy of autonomous parvoviruses employs initiation at the 3′ terminal hairpin with addition of nucleotides to form a complementary DNA strand[11] (Figure 6). The predicted double-stranded intermediate replicative forms of the B19 virus were demonstrated

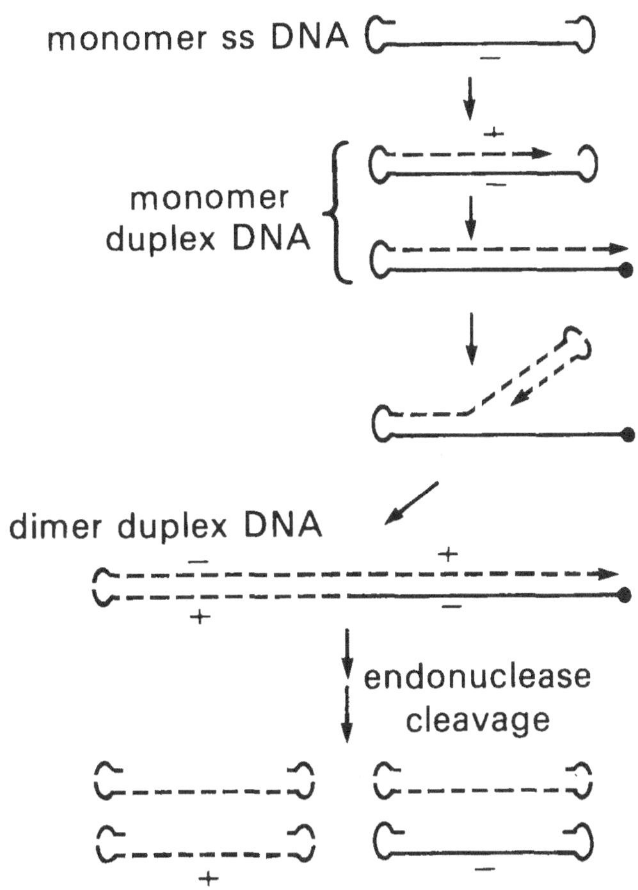

A

FIGURE 6. Replication forms of B19 virus. (A) schema of autonomous parvovirus replication; (B) Southern gel analysis of undigested virion and bone marrow nuclear DNA extracted under low salt conditions to prevent annealing of complementary single strands. (From Ozawa, K., Kurtzman, G., and Young, N., Replication of the B19 parvovirus in human erythroid bone marrow cultures, *Science*, 233, 833, 1986. With permission.)

in several ways. In the first, DNA was extracted under low salt conditions to prevent annealing in vitro of positive and negative single-stranded DNA (packaged separately in virions) (Figure 6). With this method, B19 virus-specific DNA from virions migrated as a single band of apparent molecular weight 2.5 kilobases (kb) (apparent because a single-stranded species is compared to double-stranded DNA markers). However, B19 virus-hybridizable DNA from the nuclei of infected bone marrow cells showed higher molecular weight forms corresponding to the monomer (5.2 to 5.4 kb) and dimer (10.5 kb) replicative forms as well as single-stranded B19 virus DNA. When annealed DNA was analyzed by restriction enzyme digestion and Southern transfer, extra bands were present in the lanes containing DNA from nuclei corresponding to predicted fragment sizes from the B19 virus restriction digestion map (Figure 7). Undigested virion DNA from infected sera showed a major band of 5.4 kb length, representing the annealed duplex of complementary positive and negative strands. DNA extracted from nuclei showed appropriately sized molecular weight fragments with restriction endonuclease digests of DNA using enzymes that did not cut within the viral genome (Eco R1) or cut once (Bam HI) or three times (HindIII) within the genome. Finally, the covalent

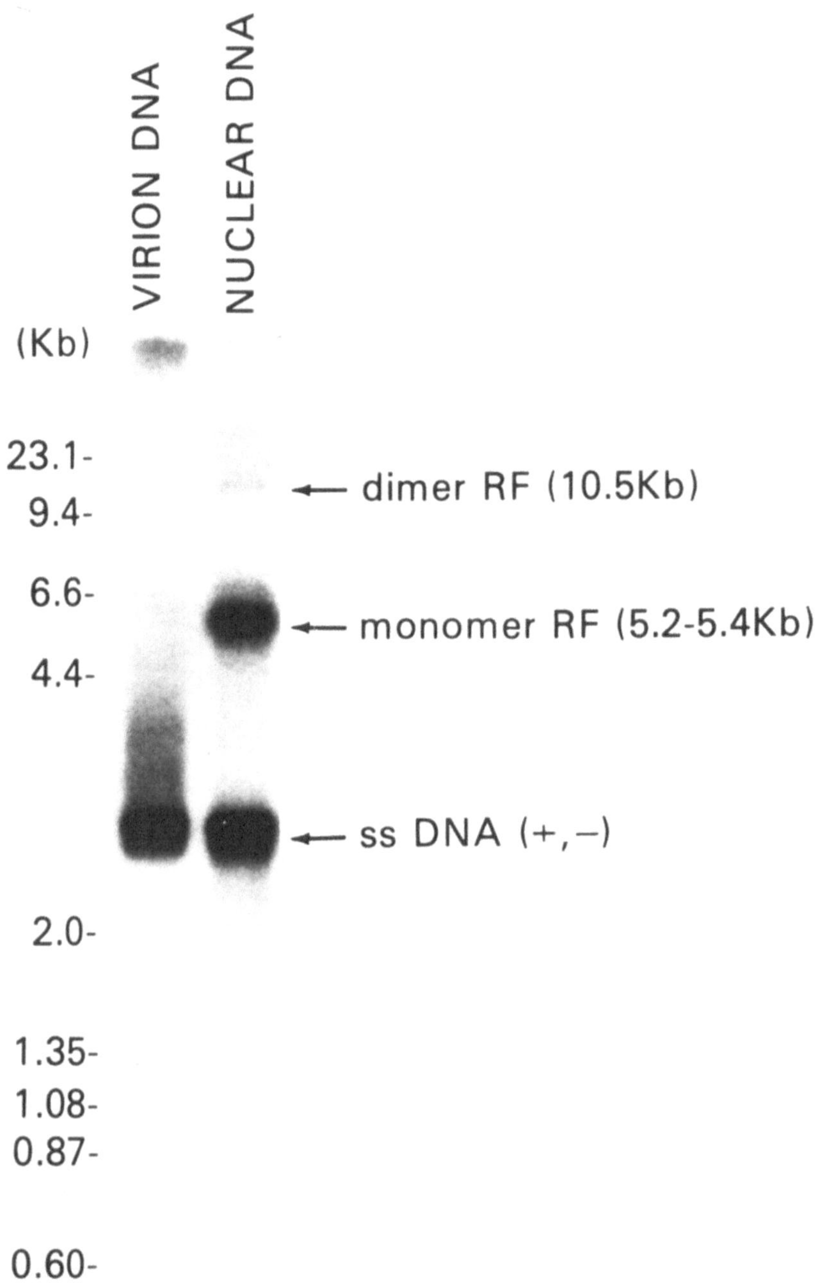

FIGURE 6B

linkage characteristic of the monomer duplex replication form could be distinguished from the annealed duplex form in a "snap back" experiment. In this method, DNA was heat denatured, then allowed to rapidly renature ("snap back") and subjected to digestion with mung bean nuclease, an enzyme that selectively digests single-stranded DNA. DNA from intact virions showed complete reduction of the duplex to single-stranded DNA under these conditions. In contrast, DNA from nuclei showed residual duplexes after denaturation that were resistant to the nuclease, confirming the covalent bond between the positive and negative strands.

Erythroid bone marrow cultures could be used to produce virus. For these experiments,

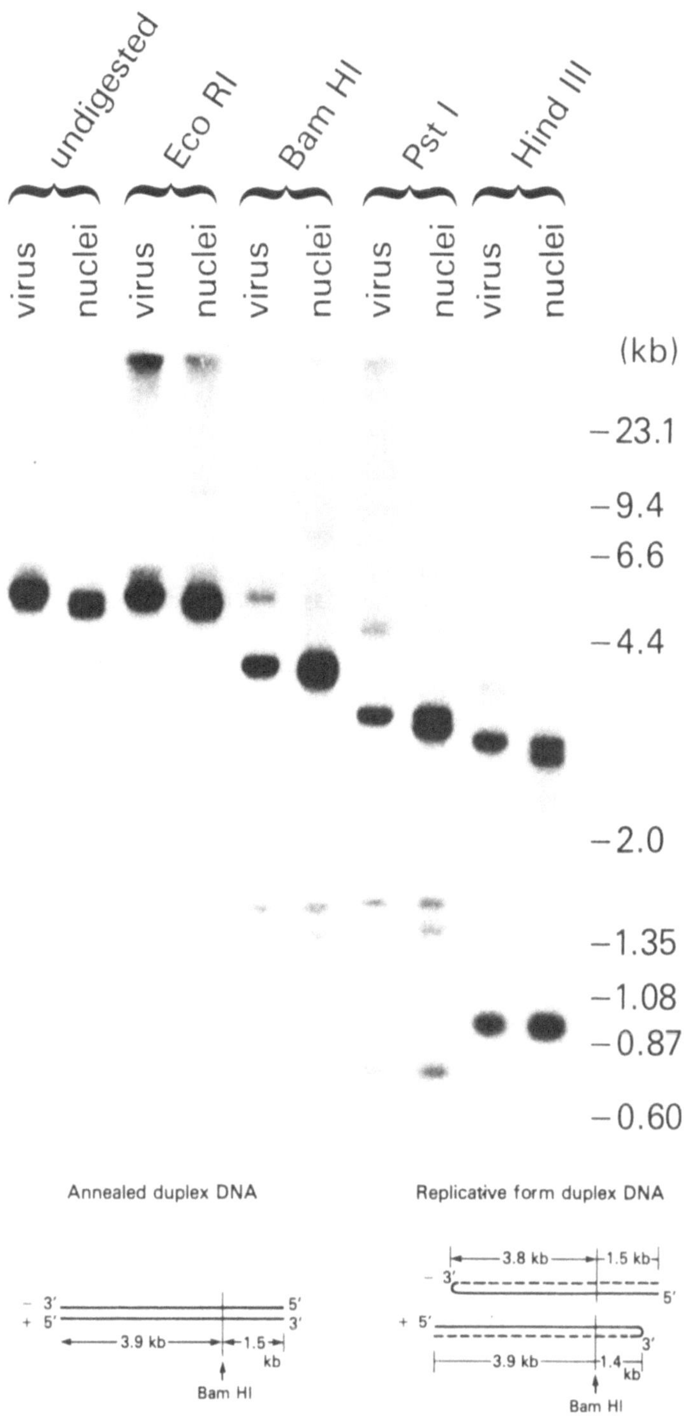

FIGURE 7. Restriction enzyme analysis of B19 virus-specific DNA from virions (sera) and nuclei (bone marrow cell cultures). Below left, schema showing generation of fragments following treatment with Bam HI, which cleaves a single site within the B19 virus genome. Correspondingly simpler and more complex patterns were obtained with other endonucleases (Eco RI, which does not cut within the genome; Hind III, which cuts at three sites).

bone marrow cells were inoculated with B19 virus containing serum and about 24 hr later washed free of the original growth medium and suspended in fresh medium. Under these conditions, viral DNA appeared in the supernatant of the cultures. At the optimal multiplicity of infection, total output of viral DNA was as high as 200-fold virus input, with the supernatant of the culture containing as much as 50-fold the viral input. The output virus usually was more infectious for fresh cultures than serum containing virus. Using immunoprecipitation or Western immunoblotting techniques and antisera (rabbit or human) to capsid proteins, parvovirus proteins of 83,000 and 60,000 kdaltons, the same sizes as detected in the sera of infected patients, appeared in the cells and supernatants of infected cultures (Figure 8). Immunofluorescence of acetone-fixed cytocentrifuge preparations of infected cultures also show capsid protein in 10 to 30% of cells, localized predominantly in the cytoplasm.

IV. BIOLOGICAL PROPERTIES OF B19 PARVOVIRUS INFERRED FROM CULTURE STUDIES

A. Physical Properties

A number of the expected physical properties of a parvovirus were demonstrated for B19 virus using infectivity in the clonal progenitor assay as an indicator of virus activity.[6] Serum containing B19 virus was resistant to heating at 60°C. Infectivity also remained after detergent treatment, consistent with the absence of an envelope, but was sensitive to proteinase treatment. Virus also was sensitive to treatment with strong acid and base.

B. Requirement for a Helper Virus

The size and restriction enzyme pattern of the cloned B19 virus suggested that the virus was of the independent genus of Parvoviridae[10] (and Chapter 2). Once the structure of the virus was determined in detail, homologies to both minute virus of mouse (an independent parvovirus) and human adeno-associated virus (dependent parvovirus) were detected by comparison of the nucleotide sequences. Cell culture studies would suggest that the virus is independent, as concurrent infection with adenovirus or herpesvirus is unlikely in all fresh bone marrow specimens, nor have other viruses been observed on electron micrographs of infected cells.[7]

C. Tissue Specificity

Infection of bone marrow cells in clonal assays inhibited erythroid colony formation but did not significantly reduce the number of colonies derived from granulocyte-macrophage progenitors.[4,5] Although in most patients with transient aplastic crisis reticulocytopenia and anemia are the characteristic peripheral blood findings, leukopenia and thrombocytopenia also occur. Similar results have been reported in normal volunteers given B19 virus intranasally.[11a] In these normal volunteers reticulocytopenia and markedly decreased CFU-E and BFU-E numbers in blood and bone marrow were found. CFU-C in the bone marrow were normal but their number was decreased in the circulation.

These results appear explicable by at least three mechanisms. First, the virus may have a relative but not absolute preference for erythroid compared to myeloid cells, not apparent with in vitro studies because of the low concentration of infecting virus necessarily employed (serum dilutions of 10^{-2} or more). In patients with underlying hemolytic anemias and consequently large amounts of virus produced, and normal volunteers inoculated with substantial amounts of virus, the virus concentration in the bone marrow may be high enough to inhibit not only red cell but also granulocyte and platelet production. A second possibility is that stem cells common to both erythroid and myeloid progenitors are infected, and that expression of virus as cytotoxicity during erythroid, myeloid, or megakaryocytic differen-

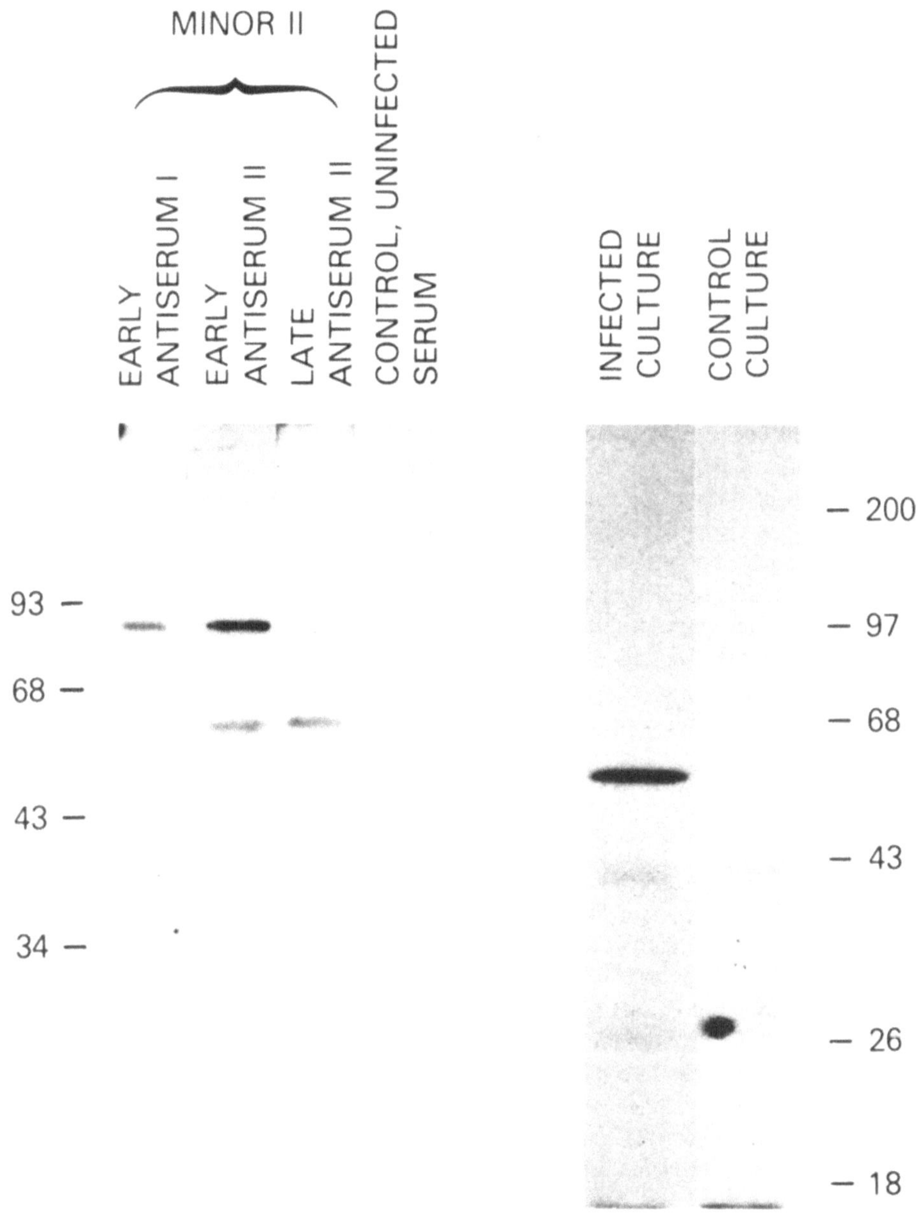

FIGURE 8. B19 virus antigens in human sera and in cells from infected cultures. On the left, immunoblots of sera containing antigens, developed with different anti-B19 virus human antisera. Capsid proteins of approximately 60 and 83 kdaltons are visualized. On the right, newly synthesized capsid protein in erythroid cell cultures was visualized by addition of ^{35}S-methionine to infected erythroid cell cultures and immunoprecipitation of viral protein with convalescent phase human antisera or rabbit antisera to capsid proteins. The 60-kdalton species is the predominant protein synthesized, although the 83-kdalton species can be visualized with more prolonged exposure of the autoradiograph. Similar data were obtained using culture supernatants.

tiation is variable. Third, depressive effects on myeloid and megakaryocytic cell lines may be a secondary phenomenon. We and others[2] have shown that interferon is a potent inhibitor of hematopoiesis in vitro, and clinical administration results in neutropenia and thrombocytopenia. Interferon production probably accompanies parvovirus infection in man. Before the identification of B19 virus as the causative agent the presence of interferon in the serum of patients during aplastic crisis was inferred from measurement of an interferon-induced enzyme; elevated levels of 2'5'-oligoadenylate synthetase were found in the peripheral blood mononuclear cells of five of seven cases, and falling levels with resolution of the clinical infection measured in four of four patients.[12]

The basis of erythroid specificity should be understood eventually from tissue culture studies. As of this writing, we know that parvovirus sequences are not found in the nuclear DNA from cultures rich in myeloid cells obtained from patients with chronic myelogenous leukemia in stable phase. Virus proliferation in bone marrow cells is dependent on the erythroid content of the cultures and partially dependent on addition of the hormone erythropoietin. Specificity may be governed either by events at the cell surface or in the nucleus. Different strains of the murine parvovirus, minute virus of mouse, vary in their ability to infect fibroblasts or lymphocytes because of factors present in the nucleus of the permissive cell type; binding to the surface of nonpermissive cells is equal to binding to permissive cells.[13] Similar binding studies using labeled B19 virus will delineate the tissue distribution of the cell surface receptor for the human parvovirus. Concurrent experiments are directed toward detecting small amounts of persistent B19 virus in nonpermissive cells like granulocyte/macrophage progenitors and lymphocytes; this finding would point to a defect in viral replication rather than viral entry as responsible for virus expression in different cell types.

V. IMPLICATIONS OF STUDIES WITH B19 VIRUS FOR VIRUS-BONE MARROW STEM CELL INTERACTION

The discovery of a specific virus that causes transient aplastic crisis has important implications for viral etiologies of other forms of bone marrow failure.[14] Aplastic anemia, in which all lineages in the marrow are depressed, often with fatal result, can follow on viral infections: non-A, non-B hepatitis, and Epstein-Barr virus infections, and, in children, upper respiratory tract infections. Laboratory studies of immune system abnormalities have suggested parallels with chronic viral infections in animals and man,[2] in particular elevated production of lymphokines like interferon and interleukin 2, increased T lymphocytes expressing Tac antigen (evidence of activation), and altered helper/suppressor lymphocyte ratios, all present in either HTLV 1 (adult T cell leukemia) or HTLV 3 (AIDS) infections. Aplastic anemia is a deficiency disease like AIDS, and a cat retrovirus, feline leukemia virus, more commonly causes bone marrow failure than it does leukemia.

Animal parvoviruses can cause chronic disease, and adeno-associated virus integrates into the genome of some human cell lines. However, parvovirus infection has not been associated with hematologic diseases other than transient aplastic crisis. The sera of patients with aplastic anemia, paroxysmal nocturnal hemoglobinuria, transient erythroblastopenia of childhood, and adult pure red cell aplasia have not shown evidence of recent infection (neither specific IgM antibody nor B19 virus antigen), and the proportion of individuals with anti-B19 virus IgG is normal.[6] Moreover, DNA analysis of cells from patients with aplastic anemia has not shown evidence of persistent B19 infection.[14a] These results do not exclude the possibility that B19 virus persists in a small fraction of the cells analyzed or that under some circumstances acute B19 virus infection might damage an hematopoietic stem cell (in contrast to its usual erythroid progenitor cell target) and lead to generalized bone marrow failure presenting after the infection had resolved. Drugs are potential candidates as agents that might extend virus host range by inducing cycling in normally quiescent stem cells;

perhaps the link between certain antibiotics like chloramphenicol and anti-inflammatories like phenylbutazone to aplastic anemia is through their use in febrile or polyarthralgic patients with bone marrow virus infections. A glaring example of misassignment of a viral etiology to a presumed toxin is the identification of glue sniffing as a cause of aplastic anemia in six cases, five of whom were children with sickle cell anemia who experienced abrupt onset of transient erythropoietic failure.[15]

The story of the discovery of the B19 virus and its link to a hematologic disease is illuminating as a model for the study of diseases caused by viruses but not normally considered "infectious" in etiology. Although the clinical characteristics of transient aplastic crisis had strongly suggested its infectious origin, the possibility that the syndrome was due to a discrete agent and not a "nonspecific" effect of many infections was not initially entertained and the role of B19 virus in vivo and in vitro met with some skepticism by hematologists. The relationship of aplastic crisis to fifth disease, obvious in the subsequent analysis of concurrent outbreaks,[16,17] was unappreciated before the identification of B19 virus. The differences in the clinical presentation of normal individuals (fifth disease) and patients with underlying hemolysis (not always apparent at the time of aplastic crisis), as between children (rash illness) and adults (arthralgia syndrome) show that one agent can, in the appropriate host, present as an exclusively hematologic, dermatologic, or rheumatic syndrome. The biologic basis for these differences remains unknown but could be due to quantitative variations in virus production in the bone marrow and the immune response mounted by the host. Without the proper assays for virus and specific antibody to the virus, not only would the syndromes appear unrelated but in vitro data showing inhibition of colony formation by sera from infected patients would be difficult to interpret. Suppression of erythropoietic colony formation by transient aplastic crisis sera only became comprehensible with the discovery of B19 virus.[18] Likely the same will occur when viruses are associated with other bone marrow diseases.

REFERENCES

1. **Metcalf, D.,** The *Hematopoetic Colony Stimulating Factors,* Elsevier, New York, 1984.
2. **Zoumbos, N., Raefsky, E., and Young, N.,** Lymphokines and hematopoiesis, *Prog. Hematol.,* 16, 201, 1985.
3. **Dexter, T. M., Simmons, P., Purnell, R. A., Spooncer, E., and Schofield, R.,** The regulation of hemopoietic cell development by the stromal cell environment and diffusible regulatory molecules, in *Aplastic Anemia: Stem Cell Biology and Advances in Treatment,* Young, N., Levine, A. S., and Humphries, R. K., Eds., Alan R. Liss, New York, 1984, 13.
4. **Mortimer, P. P., Humphries, R. K., Moore, J. G., Purcell, R. H., and Young, N. S.,** A human parvovirus-like virus inhibits haematopoietic colony formation in vitro, *Nature (London),* 302, 426, 1983.
5. **Duncan, J., Potter, C. G., Cappellini, M. D., Kurtz, J. B., Anderson, M. J., and Weatherall, D. J.,** Aplastic crisis due to parvovirus infection in pyruvate kinase deficiency, *Lancet,* 2, 11, 1983.
6. **Young, N. S., Mortimer, P. P., Moore, J. G., and Humphries, R. K.,** Characterization of a virus that causes transient aplastic crisis, *J. Clin. Invest.,* 73, 224, 1984.
7. **Young, N. S., Harrison, M., Moore, J. G., Mortimer, P. P., and Humphries, R. K.,** Direct demonstration of the human parvovirus in erythroid progenitor cells infected in vitro, *J. Clin. Invest.,* 74, 2024, 1984.
8. **Cohen, B., Mortimer, P. P., and Pereira, M. S.,** Diagnostic assays with monoclonal antibodies for the human serum parvovirus-like virus, *J. Hyg.,* 91, 113, 1983.
9. **Ozawa, K., Kurtzman, G., and Young, N.,** Replication of the B19 parvovirus in human erythroid bone marrow cultures, *Science,* 233, 833, 1986.
10. **Cotmore, S. F. and Tattersall, P.,** Characterization and molecular cloning of a human parvovirus genome, *Science,* 226, 1161, 1984.
11. **Hauswith, W. W.,** Autonomous parvovirus DNA structure and replication, in *The Parvoviruses,* Berns, K. I., Ed., Plenum Press, New York, 1983, 129.

11a. **Potter, C.,** personal communication.

12. **Williams, B. R. G., Read, S. E., Freedman, M. H., Carver, D. H., and Gelfand, E. W.,** The assay of 2-5A synthetase as an indicator or interferon activity and virus infection *in vivo*, in *Interferons*, Merigan, T. C. and Friedman, R. M., Eds., Academic Press, New York, 1982, 253.

13. **Spalholz, B. A. and Tattersall, P.,** Interaction of minute virus of mice with differentiated cells; strain-dependent target cell specificity is mediated by intracellular factors, *J. Virol.*, 46, 937, 1983.

14. **Young, N. and Mortimer, P.,** Viruses and bone marrow failure, *Blood*, 63, 729, 1984.

14a. **Kurtzman, G. and Young, N.,** unpublished observations.

15. **Powars, D.,** Aplastic anemia secondary to glue sniffing, *N. Engl. J. Med.*, 273, 700, 1965.

16. **Chorba, T., Coccia, P., Holman, R. C., Tattersall, P., Anderson, L. J., Sudman, J., Young, N. S., Kurczynski, E., Moir, R., Lawrence, D. N., Jason, J. M., and Evatt, B.,** Parvovirus B19 — the cause of aplastic crisis and erythema infectiosum (fifth disease), *J. Infect. Dis.*, 154, 383, 1986.

17. **Saarinen, U. M., Chorba, T. L., Tattersall, P., Young, N. S., Anderson, L. J., and Coccia, P. F.,** Human parvovirus B19 induced epidemic red cell aplasia in patients with hereditary hemolytic anemia, *Blood*, 67, 1411, 1986.

18. **Lutton, J. D., Schmalzer, E. A., Rao, S. P., and Levere, R. D.,** Erythroid colony studies on sickle cell anemia in hypoproliferative crisis, *Am. J. Hematol.*, 8, 15, 1980.

Chapter 9

B19 VIRUS INFECTIONS IN PREGNANCY

J. R. Pattison

TABLE OF CONTENTS

I. INTRODUCTION

In Chapter 3 details are given of the ability of some animal parvoviruses to cause damaging intrauterine infections. In general the most serious effects occur as a consequence of first and second trimester infections which are often associated with intrauterine death in hamsters, pigs, and cattle. Infections in the third trimester tend to result in distinct clinical illnesses in the newborn viz. cerebellar ataxia (cats) and myocarditis (dogs). These intrauterine infections are assumed to occur as a consequence of viremia in the pregnant animal with the developing embryo offering a variety of dividing and differentiating tissues for parvovirus replication. In human medicine it is clear that the majority of spontaneous abortions, intrauterine deaths, and congenital abnormalities are without satisfactory explanation. In acute B19 virus infection there is an intense viremia with every opportunity for seeding of the rapidly dividing tissues of the fetus should infection occur during pregnancy. Therefore it is pertinent to wonder whether B19 infection has any adverse effect on pregnancy.

This above possibility was raised long before the etiological agent of erythema infectiosum was known to be a parvovirus. One can only assume that the question of teratogenicity was first considered because of the similarity between erythema infectiosum and rubella. A very large outbreak of erythema infectiosum in Washington, D.C., was documented by Ager and colleagues.[1] The outbreak lasted from December 1961 to June 1962 and there were 667 reported cases of which 364 were studied. There were only seven pregnant women among these cases and the outcome of their pregnancies is not stated in the original paper. However, members of the medical profession in the locality were asked to note carefully any birth defects during 1962 and 1963. No increase was noted compared with annual figures from 1955 to 1964, in fact both were comparatively low incidence years. The authors concluded that there was no evidence of teratogenicity associated with erythema infectiosum during pregnancy. Cramp and Armstrong[2] also found no evidence of teratogenicity in their observations on an epidemic in a town in the west of England. They followed up all the women in the area who were in the first trimester of pregnancy during the epidemic and its incubation period. There were 29 successful pregnancies and 3 spontaneous abortions during this time. In the same period of the previous year there were 25 successful pregnancies and again 3 spontaneous abortions. One member of the control group had spina bifida and another had deafness as a result of first trimester rubella. There were no significant abnormalities in the babies born to mothers whose pregnancies coincided with the oubreak of erythema infectiosum. Thus early studies gave no support to the view that the causative virus of erythema infectiosum damages the fetus.

II. CONFIRMED B19 VIRUS INFECTIONS COMPLICATING PREGNANCY

In 1984 two reports[3,4] appeared describing intrauterine B19 infection in association with an unfavorable outcome of pregnancy. In one pregnancy an asymptomatic 24-year-old primigravida living in the northeast of Scotland had a routine serum α-fetoprotein measurement at 17 weeks gestation.[3] This was abnormal; an ultrasound scan demonstrated hydrops fetalis and the pregnancy was terminated at 19 weeks gestation. At necropsy generalized edema of a male fetus of about 19 weeks gestation and effusions of the pleural, pericardial, and peritoneal cavities were found. Histologically, there was a marked leukoerythroblastic response in the liver, spleen, and kidneys. The fetal circulation and placenta contained many nucleated red blood cell precursors and the liver had extensive iron deposits suggesting hemolysis. Laboratory investigations revealed a normal karyotype, no blood group incompatibility, and no evidence of recent infection with many agents including cytomegalovirus and rubella virus. However, serology indicated a seroconversion for B19 virus-specific IgM and IgG antibody between the 7th and 16th weeks of gestation. Although it is difficult to be more precise than this the pattern of the changes in amounts of B19 virus-specific IgM and IgG antibody suggest infection between the 13th and 15th week of gestation. Parvovirus DNA was detected by dot-blot hybridization in the placenta and in fetal kidney, adrenals, heart, thymus, and liver.

The second pregnancy[4] reported to be complicated by intrauterine parvovirus infection, was that of a 35-year-old para 0 + 1 in southeast England. The early part of the pregnancy was uneventful and serum and amniotic α-fetoprotein levels were normal at 16 weeks gestation. The patient reported a flu-like illness at 39 weeks and 8 days later spontaneous labor started. On admission the fetal heartbeat could not be recorded and intrauterine death was confirmed by ultrasound. A macerated, stillborn girl weighing approximately 4 kg was delivered and severe ascites were the only obvious abnormality. Necropsy revealed that the ascites were due to meconium ileus and peritonitis, but there was no evidence of cystic fibrosis which the authors state would be the most likely cause of the abdominal findings. Serological tests on maternal and fetal blood showed no evidence of toxoplasmosis, rubella, cytomegalovirus, or herpes simplex virus infection. However, both blood samples contained B19 virus-specific IgM indicating recent maternal infection and transplacental infection of the fetus.

These two cases raised the possibility that B19 infection in pregnancy is associated with fetal loss and is a possible cause of hydrops fetalis. A further isolated case of midterm fetal loss with hydrops has been reported.[5] A 27-year-old woman with two normal children living in the west of England presented with a rubella-like rash at 15 weeks gestation. Serologically this proved to be due to B19 virus infection. The rash persisted with mild fever for several weeks and stiffness of the knees and hands developed. She remained unwell and complained of lower abdominal pain until 25 weeks gestation when fetal movements ceased and a stillborn hydropic fetus was delivered. B19 virus DNA was detected in placenta and fetal liver, lung, brain, and adrenal. The placenta was shown to contain a large number of parvovirus particles and small numbers were present in the fetal tissues. Maternal serum was free of virus at the time of the abortion, indicating that the virus detected in the fetus was not due to persistent maternal infection.

Although there is some uniformity in the individual cases described above, it is possible, with such small numbers, that one is observing the coincidence of two unrelated events. In one case[4] the authors state that B19 virus infection was epidemic in the area at the time of the stillbirth. The second case[3] was also identified during an epidemic of B19 virus-associated fifth disease in the local community. In a later publication it became clear that it was one of six cases of serologically proved B19 virus infection occurring during the first half of pregnancy.[6] Four were delivered of apparently normal infants at term but the other two, including the one described above, resulted in midtrimester abortions with remarkably similar features in the fetuses. Both had hydrops fetalis of nonimmunological origin, both had a severe leukoerythroblastic reaction, and both had extensive iron deposition in the liver. In both cases, B19 virus DNA was detected in several fetal organs. Not only is there a striking similarity between the two cases of fetal loss in this series, but the overall incidence (two out of six) is high. Again the numbers studied are small, but similar small studies tend to give a consistent picture. Eight patients with B19 virus infection in pregnancy (4 to 17 weeks gestation) were identified among a group of individuals with erythematous rashes which were shown not to be due to rubella.[7] Three of these had spontaneous abortions, 1, 3, and 6 weeks after the rash; one had a therapeutic abortion for unrelated reasons; three delivered apparently normal infants, and two of these had B19 virus-specific IgG at 12 and 16 months of age. Thus in two studies there are fetal loss rates of 33% (two of six) and 37.5% (three of eight). Similar rates have been found in small studies in France,[8] Sweden, and Iceland.[8a]

Another interesting finding that has emerged from two independent studies is the relative infrequency of confirmed B19 virus infection in pregnant women compared with nonpregnant individuals. In the study referred to above[7] of 367 adults with rubella-like rashes which were known not to be due to rubella virus infection, B19 infection was diagnosed in 8 of 222 (3.5%) pregnant women and 22 of 145 (15.2%) other patients of both sexes. In another study of rubelliform rashes 164 cases were investigated for rubella and B19 virus infection.[9] A total of 39 (24%) were shown to be due to rubella virus infection and 13 (8%) to B19 virus infection. Of all the cases investigated, 98 were women in the childbearing years and 31 (32%) were pregnant. Twenty four of the confirmed rubella cases were in this category and nine (38%) were pregnant. Seven of the confirmed B19 virus infections were in the same category but none of these were pregnant. Thus, it seems that B19 virus infection does not occur with the expected frequency in pregnant women. The reasons for this are not known at present but the possibilities are (1) that pregnancy has a modifying effect on the clinical features of B19 infection or (2) that the virus infection leads to early fetal loss, i.e., before the pregnancy is confirmed.

III. B19 INFECTION AND CONGENITAL ABNORMALITIES

Clearly, B19 virus infection does not always have an unfavorable effect on pregnancy. Indeed, the majority of established pregnancies complicated by this infection go to term and result in the birth of infants with no obvious abnormality. The lack of association between B19 virus infection and obvious birth defects has been confirmed by examining sera from abnormal infants. Stored sera taken mostly within the first month of life from infants with birth defects in Manchester, England and Bern, Switzerland were tested for B19 virus-specific IgM and for the virus.[7] Neither specific IgM antibody nor viral antigen were detected in any of the sera. However, the authors point out that diagnosis of intrauterine viral infection by the detection of specific IgM antibody in cord blood or serum taken shortly after birth is based on the model of congenital rubella and cytomegalovirus (CMV) infection. Both are persistent infections and the causative virus can be isolated at birth and beyond. Intrauterine B19 virus infection may be an acute event with eventual elimination of virus. This may not result in the presence of specific IgM at birth and it may be necessary to test sera taken at 12 months of age for specific IgG antibody in order to diagnose intrauterine B19 virus infection. In theory this might only apply if the infection occurred after the fetus had become immunocompetent at about 18 to 20 weeks and much organogenesis would have taken place before that stage of gestation. However IgG antibody (presumably persistent) at 12 and 16 months of age has been detected in two infants born to mothers who had B19 infection between 4 and 17 weeks gestation.

IV. CONCLUSION

At the beginning of this chapter it was predicted that the intense viremia associated with acute B19 virus infection would lead to fetal infection should it occur during pregnancy. Clearly from the studies outlined above, this is the case and both placental and fetal tissues are permissive for B19 virus replication. However, at the time of writing, the clinical consequences of this intrauterine infection cannot be stated with certainty. The numbers involved are small and the studies have been retrospective. Nevertheless, the evidence suggests that intrauterine B19 infection is a cause of fetal loss. This seems to be the case at whatever stage of pregnancy the infection occurs, from the first to the last month. The adverse effects do not seem to occur simultaneously with the acute B19 virus infection, rather there is an interval of 1 to 10 weeks between infection and fetal loss. Early fetal loss can only be classified as a spontaneous abortion but at later stages of gestation the infection may well be a cause of nonimmunological hydrops fetalis.

Early clinical studies related to outbreaks of erythema infectiosum did not provide evidence for a teratogenic effect of the causative virus. Recent virological studies have confirmed this view and there remains no evidence for a defined birth defect or even a variety of them as a consequence of intrauterine B19 infection. Nevertheless, late onset defects have not yet been excluded and the search for specific IgM at birth may not be a sensitive indicator of intrauterine B19 infection. There is no doubt that carefully controlled prospective studies are required to determine the effect of B19 virus infection on pregnancy and the developing fetus and that infants born as a result of pregnancies complicated by B19 infection must be followed up, clinically and virologically, for some years.

REFERENCES

1. **Ager, E. A., Chin, T. D. T., and Poland, J. D.,** Epidemic erythema infectiosum, *N. Engl. J. Med.*, 275, 1326, 1966.
2. **Cramp, H. E. and Armstrong, B. D. J.,** Erythema infectiosum: no evidence of teratogenicity, *Br. Med. J.*, 2, 1031, 1977.
3. **Brown, T., Anand, A., Ritchie, L. D., Clewley, J. P., and Reid, T. M. S.,** Intrauterine parvovirus infection associated with hydrops fetalis, *Lancet*, 2, 1033, 1984.
4. **Knott, P. D., Welply, G. A. C., and Anderson, M. J.,** Serologically proved intrauterine infection with parvovirus, *Br. Med. J.*, 289, 1660, 1984.
5. **Bond, P. R., Caul, E. O., Usher, J., Cohen, B. J., Clewley, J. P., and Field, A. M.,** Intrauterine infection with human parvovirus, *Lancet*, 1, 448, 1986.
6. **Gray, E. S., Anand, A., and Brown, T.,** Parvovirus infections in pregnancy, *Lancet*, 1, 208, 1986.
7. **Mortimer, P. P., Cohen, B. J., Buckley, M. M., Cradock-Watson, J. E., Ridehalgh, M. K. S., Burkhardt, F., and Schilt, U.,** Human parvovirus and the fetus, *Lancet*, 2, 1012, 1985.
8. **Lefrere, J. J., Dumez, Y., Courouce, A. M., and Deschenne, G.,** Intrauterine infection with human parvovirus, *Lancet*, 1, 449, 1986.

8a. **Anderson, M. J.,** personal communication.

9. **Anderson, M. J., Kidd, I. M., and Morgan-Capner, P.,** Human parvovirus and rubella-like illness, *Lancet*, 2, 663, 1985.

Chapter 10

SMALL ROUND HUMAN FECAL VIRUSES

E. Owen Caul

TABLE OF CONTENTS

I. SMALL ROUND HUMAN FECAL VIRUSES

The study of nonbacterial gastroenteritis came of age in the early 1970s after many fruitless years spent in the search for an etiological agent. Virological aspects of this syndrome were studied in the 1950s and 1960s, using cell culture techniques, with very little reward for much painstaking laboratory work. With the successful development of organ culture techniques for the study of viruses[1] a possible approach to the study of viral causes of gastroenteritis became available. Although initial aspirations were high that organ cultures derived from human fetal intestine might provide the means of detecting fastidious enteric viruses, only limited success was achieved. Replication of a number of conventional human fecal viruses, which could also be propagated in cell culture, occurred in these intestinal organ cultures,[2-4] but not the elusive agents of nonbacterial gastroenteritis. The discovery of these agents awaited the application of newer technology and alternative approaches to the problem.

Two groups of laboratory workers in America and England examined by direct electron microscopy fecal suspensions from cases of nonbacterial gastroenteritis. In 1972 the American group published the results of their studies[5] describing a 27-nm small round virus particle which was detected in stools from an outbreak of nonbacterial gastroenteritis which had occurred in the town of Norwalk, in Ohio. Shortly after this description of the Norwalk agent, the English workers reported the detection of a 22-nm virus particle (Wollan agent) in stools from patients with a similar clinical syndrome to that described in the American studies.[6] Both groups of workers described the small round viruses that they had observed as parvovirus-like. Even at this early stage it was apparent that these two small round viruses, which had been reported separately on both sides of the Atlantic, differed considerably in morphology.

Electron microscopy then became fashionable for the study of nonbacterial gastroenteritis and as a result of the application of this "new" technology to the direct examination of stools, numerous viruses have been described. Many of these new human viruses have become known under the umbrella term of small round viruses. This term describes those virus or virus-like particles measuring less than 40 nm in diameter, some of which have a distinctive morphology allowing definitive identification but others have been described as parvovirus/picornavirus-like particles. A summary of the small round viruses described to date is shown in Table 1. Most were known by the geographical location in which they were found, largely because detailed knowledge on their taxonomic status was not available. This lack of information on the basic properties of the small round viruses has led to many being incorrectly classified as parvovirus-like and this in turn has led to considerable confusion in the literature — a situation which remains to this day. As a result of the unsatisfactory states regarding virus nomenclature, a proposal was made for the interim classification of small round viruses.[7] The classification scheme was based on an electron microscopical study which was carried out on many but not all of the small round viruses described in the literature. An integral and essential factor of the study was that the viruses were examined in the absence of any added antibody which is known to mask surface structure present on virus particles. This interim scheme divides small round viruses into those that possessed a surface structure (structured viruses) and those that lacked any discernible surface features (featureless viruses) and is shown in Table 2. It has alleviated some of the problems which existed and enabled useful epidemiological data to be collected in the U.K. Since this scheme was proposed other fundamental properties of some of the small round viruses have been established which further support the proposed classification scheme. In the light of this new information on small round viruses in human feces, a brief review of their etiological role and a reappraisal of their classification is proposed with particular emphasis on which, if any, of these small round viruses are likely to be parvovirus candidates.

Table 1
HISTORICAL RECORD OF SMALL ROUND VIRUSES

Name	Year reported	Derivation of name	Country	Ref.
Norwalk virus	1972	Geographical	America	5
Wollan agent	1973	Geographical	England	6
Astrovirus	1975	Descriptive	England	8
Calicivirus	1976	Descriptive	England	31
Hawaii virus	1977	Geographical	America	60
Montgomery County	1977	Geographical	America	60
Ditchling agent	1977	Geographical	England	59
Cockle agent	1977	Epidemiological	England	73
Paramatta agent	1978	Geographical	Australia	89
Otofuke agent	1979	Geographical	Japan	78
Sapporo agent	1981	Geographical	Japan	80
Marin County agent	1981	Geographical	America	81
Snow Mountain agent	1982	Geographical	America	82

Table 2
DIVISION OF SMALL ROUND VIRUSES ACCORDING TO MORPHOLOGY

Morphology	Virus strains
Structured particles; classical surface structure or amorphous surface structure lacking geometrical symmetry	Classical structure Astrovirus Calicivirus Amorphous structure Norwalk virus Hawaii virus Montgomery County virus Norwalk-like viruses[a]
Featureless particles; no resolvable surface structure and a sharply delineated outer edge	Wollan agent Ditchling agent Cockle agent Parramata agent Hepatitis A (reference picornavirus strain) Coxsackie B5 (reference picornavirus strain) Mink enteritis virus (reference parvovirus strain)

[a] Virus particles morphologically indistinguishable from Norwalk virus.

Adapted and modified from Caul and Appleton.[7]

II. STRUCTURED VIRUSES

A. Astroviruses

During the course of an electron microscopical study of infants feces, Madeley and Cosgrove[8] described small round, approximately 28 nm virus particles, possessing a distinct five- or six-pointed surface star (Figure 1). This unique surface star is usually visualized on 10 to 20% of the virus particles present in any sample[9,10] and dictates the accurate identification of the virus. Commonly, astroviruses are present in large numbers, sometimes as semicrystalline arrays or as small aggregates. No definitive evidence has been produced to show that these naturally occurring virus aggregates are the result of antigen/antibody complexes. The overall impression when visualizing astroviruses is that they are extremely uniform in size (28 $\pm$ 1.0 nm), possessing a smooth entire edge and a surface star —

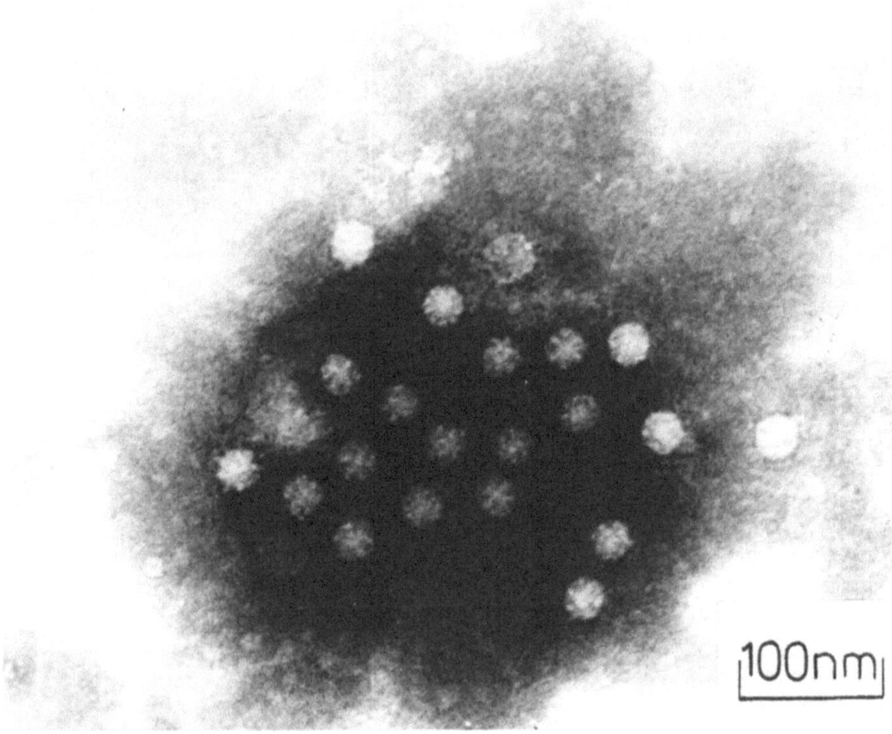

FIGURE 1. The appearance of astrovirus particles in feces after ammonium sulfate precipitation. Note the surface star on some of these particles which allows a definitive identification. (Stained with 1.5% phosphotungstic acid pH 6.5.)

features which are diagnostic of astroviruses and allow differentiation from caliciviruses with which they have been confused.[11,12] Under ideal conditions of specimen preparation, projections measuring approximately 5 nm in length, can be visualized as evenly spaced rod-like structures in close association with the surface of astrovirus particles.[13] A detailed description of the distinguishing features of astroviruses and caliciviruses has been published.[13] Astroviruses are unstable if they are pelleted from cesium chloride gradients but purification and stability in tartrate/glycerol gradients has been demonstrated where they band at a density of 1.32 g/cm^3.[14] To date, no information is available on the nucleic acid of human astroviruses, but studies on a lamb astrovirus demonstrated a single-stranded ribonucleic acid (RNA) genome.[15] On the basis of the biochemical data obtained by these workers they suggested that astroviruses should be considered as a new group with a polypeptide composition intermediate between picornaviruses and caliciviruses.[15]

Since the first description of astroviruses in human feces by Madeley and Cosgrove in 1975, many workers have recorded identical particles. The virus has been detected in approximately 5% of fecal samples from symptomatic infants and children[10] with a somewhat higher incidence in neonates often in the absence of any symptoms.[8,16-19] Although asymptomatic infection is common, clinically recognizable disease usually occurs sporadically in neonates and children. Nosocomial spread has been documented and outbreaks of astrovirus gastroenteritis in infants and children have been described.[9,17,20-22] In one of these outbreaks the virus spread to ward staff, indicating that susceptibility is not confined to neonates and children.[20] In a large pediatric outbreak of astrovirus gastroenteritis in a nursery in Japan over 50% of the children presented with symptoms.[21] A close temporal association of astrovirus excretion with the acute stage of gastroenteritis has been documented[9] and this in association with rising antibody levels and specific IgM responses[9,21] suggested these viruses were enteric pathogens. Recent studies in adult volunteers have confirmed that astroviruses are pathogens when a fecal filtrate containing astroviruses produced diarrhea and vomiting in one of eight volunteers.[23] Passage of a fecal filtrate from this ill volunteer resulted in subclinical infection with excretion of virus in two of nine volunteers. In all, seroconversions occurred in 13 out of 16 volunteers studied. At present it is assumed that human astroviruses infect and replicate within the mucosal cells of the small intestine but no direct evidence exists to support this view. However, studies in animals present conclusive evidence that the lamb astrovirus replicates within mucosal cells of the small intestine resulting in a mild villous atrophy.[24]

Limited sero-epidemiological data is available on the acquisition of astrovirus antibody in the community. A study carried out in Oxford showed that 7% of 6- to 12-month-old infants had specific antibody and this increased progressively in 1- to 4-year-old children to a level of 75% in 5 to 10-year-olds, which was very similar to the incidence in young adults.[25] This study is of particular interest in view of the low frequency of detecting astroviruses in fecal samples by electron microscopy. Several possible explanations may be relevant to these findings. Asymptomatic infection may be very common, or alternatively, infection may give rise to such mild gastrointestinal upsets that medical intervention is not warranted. The latter suggestion is supported by the clinical description of an astrovirus outbreak in Japan[21] where only 1 of 46 ill children required medical care. Another possibility is that astroviruses are underdiagnosed in the laboratory because the characteristic surface star is not demonstrated.[7]

The sero-epidemiological studies carried out by the Oxford group were facilitated by the development of techniques for propagating astroviruses in cell culture.[26,27] Using immunofluorescence and immune electron microscopy these workers have established that five astrovirus serotypes exist.[28] The ability to propagate serially at least some human astroviruses will facilitate the biochemical characterization of this group of uniquely structured small round viruses[13] and will confirm their RNA genome and polypeptide composition,[15] thus allowing a more definitive taxonomic status.

B. Caliciviruses

Caliciviruses are a morphologically unique group of viruses which have been removed recently from the picornavirus group to form the new family Caliciviridae.[29] Classical particles (Figure 2) showing fivefold symmetry, possess a central, stained-filled cup surrounded by six peripheral cups giving rise to the so called "Star of David" appearance.[13] The appearance of caliciviruses in the electron microscope will vary according to the orientation of the particle and virus particles showing two- or threefold symmetry are more commonly recognized. Most workers will accept a virus as a calicivirus if particles show the classical "Star of David" appearance. Caliciviruses have a buoyant density of 1.38 to 1.40 g/cm^3 in cesium chloride[30] and a particle diameter ranging from 31 to 35 nm.[10,13] Unlike astroviruses which possess a smooth entire edge, the ragged outline of caliciviruses makes accurate measurements difficult and explains the wide range of particle diameters quoted in the literature.

The first report that a human calicivirus may be associated with nonbacterial gastroenteritis came in 1976 when Madeley and Cosgrove[31] described typical particles in the feces of children. Shortly after this report Flewett and Davies[32] reported caliciviruses in the small intestine of a child who had died from gastroenteritis. However, an etiological role for this virus in nonbacterial gastroenteritis could not be established in these early studies because the virus was present in the stools of both healthy and symptomatic children.[31] The first convincing report that caliciviruses can give rise to gastroenteritis came from studies of winter vomiting disease in London.[33,34] These workers, using the combined approach of

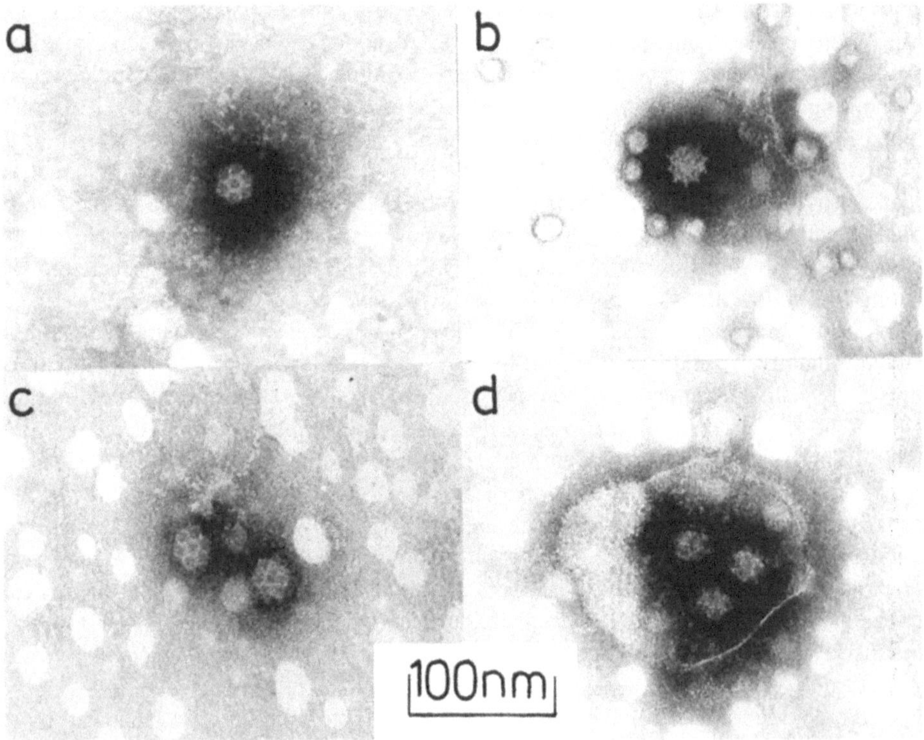

FIGURE 2. Morphological variation of caliciviruses present in a single stool sample after ammonium sulfate precipitation. (a) Classical calicivirus showing a central stain-filled cup surrounded by six peripheral stain-filled cups (threefold symmetry); (b) Calicivirus particle showing fivefold symmetry. Note the ten regular spaced points around the edge of the particle due to its different orientation; (c) Two calicivirus particles showing surface stain-filled cups. Note only the one particle shows the classical threefold symmetry; (d) Calicivirus particles surrounded by a membrane. Note the amorphous appearance of the particles and the lack of surface stain-filled cups. (Stained with 1.5% phosphotungstic acid pH 6.5.)

direct electron microscopical examination of fecal samples and immune electron microscopy for serological evidence, concluded that caliciviruses were a cause of nonbacterial gastroenteritis. This report was followed by the description of outbreaks of infantile gastroenteritis in Japan,[35,36] where seroconversions and IgM responses to a calicivirus were demonstrated,[36] and a further outbreak in Japanese schoolchildren.[37] Further reports from England,[38] Japan,[30] and Canada[39] also suggested a causal role in gastroenteritis where nosocomial and sporadic cases were reported. It appears from these reports that symptomatic infection is more frequent in children than in neonates. To date only one outbreak of calicivirus gastroenteritis in the elderly has been reported.[40]

A sero-epidemiological study in Japan using immune electron microscopy suggests that calicivirus infection is widespread in the community in Sapporo.[41] In this study 65% of infants between 6 and 23 months and 90% of schoolchildren and adults were shown to be sero-positive. The high prevalence rates of calicivirus antibody in the Japanese studies are particularly interesting. If similar high antibody incidences were to be demonstrated in the U.K., it would contrast with the fact that caliciviruses are uncommonly detected in the stools. Laboratory data on the frequency of calicivirus detection by electron microscopy in Japan in sporadic cases of gastroenteritis is not available at present. This situation may be very similar to astrovirus infections where a high incidence of subclinical infections may occur or infection results in such mild gastrointestinal symptoms that medical care is not warranted. Clinical observation of calicivirus gastroenteritis suggests that symptoms are milder than those due to rotaviruses.[42] The recent development of a solid-phase radioimmunoassay for the detection of caliciviruses in stools[43] will further increase our knowledge on the role of these viruses in gastroenteritis and aid sero-epidemiological studies. Cubitt et al.[40] have recently demonstrated by immune electron microscopy that at least two calicivirus serotypes exist.

In common with most of the small round viruses, progress in calicivirus research has been hampered by the lack of a suitable laboratory system for propagating the virus. Recently one strain of human calicivirus has been propagated in human embryo kidney cells.[44] Virus multiplication was demonstrated in the presence of actinomycin D and on the basis of radiolabeling experiments with [^{3}H] uridine these workers concluded that the virus had an RNA genome. Unfortunately these workers were unable to demonstrate that the virus had a major polypeptide of 66,000 mol wt which in conjunction with its RNA genome and typical morphology would have allowed definitive classification. Nevertheless, in view of the unique morphology of caliciviruses it is extremely unlikely that this human enteric virus has been incorrectly classified.

C. Norwalk Virus

The town of Norwalk, Ohio set the scene for the first definitive studies of viral causes of nonbacterial gastroenteritis. An outbreak of winter vomiting disease[45] occurred in an elementary school in Norwalk which spread to family contacts. A high secondary attack rate was noted and this high secondary spread remains a feature of these infections today. Fecal material derived from an adult secondary contact with symptoms was shown to reproduce an identical illness in adult volunteers.[46] Subsequently a small round virus measuring approximately 27 nm in diameter was identified in fecal specimens derived from volunteers involved in these transmission experiments.[5] This small round virus was termed the Norwalk agent (Figure 3) and on the basis of its resistance to ether, acid, and heat, its size and buoyant density in cesium chloride, it was considered to be parvovirus-like.[5,47,48] Since these early descriptions of the Norwalk virus extensive volunteer studies by the American workers have established this group of viruses as important etiological agents of gastroenteritis.[40,49-51] Infection with Norwalk virus is associated with an acute, mild, self-limiting infection.[52] Volunteer studies have demonstrated a mild villous atrophy of the small intestine[51-53] but virus replication in mucosal cells has not been demonstrated.

Sero-epidemiological studies have shown that Norwalk virus infection occurs on a worldwide scale[54-56] with acquisition of antibody occurring more commonly in younger age groups in underdeveloped countries than in developed countries.[56] The development of a radioimmunoassay (RIA) by the American workers assisted them in their serological surveys but the scarcity of specific reagents has greatly hampered the development of similar assays in research laboratories elsewhere in the world.

The problems associated with the relative insensitivity of the electron microscope highlights the need for alternative assays for the detection of both antigen and antibody. The lack of a suitable cell culture system has hampered the production of specific antisera which would resolve problems associated with assays based on human sera from infected patients. Attempts to reproduce the disease in laboratory animals have proved disappointing[57] but inoculation of chimpanzees with a fecal filtrate containing Norwalk virus resulted in excretion of virus and antibody conversion[58] in the absence of symptoms. Sera from these chimpanzees have been useful in the development of an RIA but the scarcity of even these reagents has limited their general use to the American studies. It is interesting that a preinoculation serum from this chimpanzee reacted with the unrelated Ditchling agent (discussed later) by immune electron microscopy.[59] These results require confirmation but highlight the need for caution in the interpretation of sero-epidemiological studies where specificity of the reagents is in doubt.

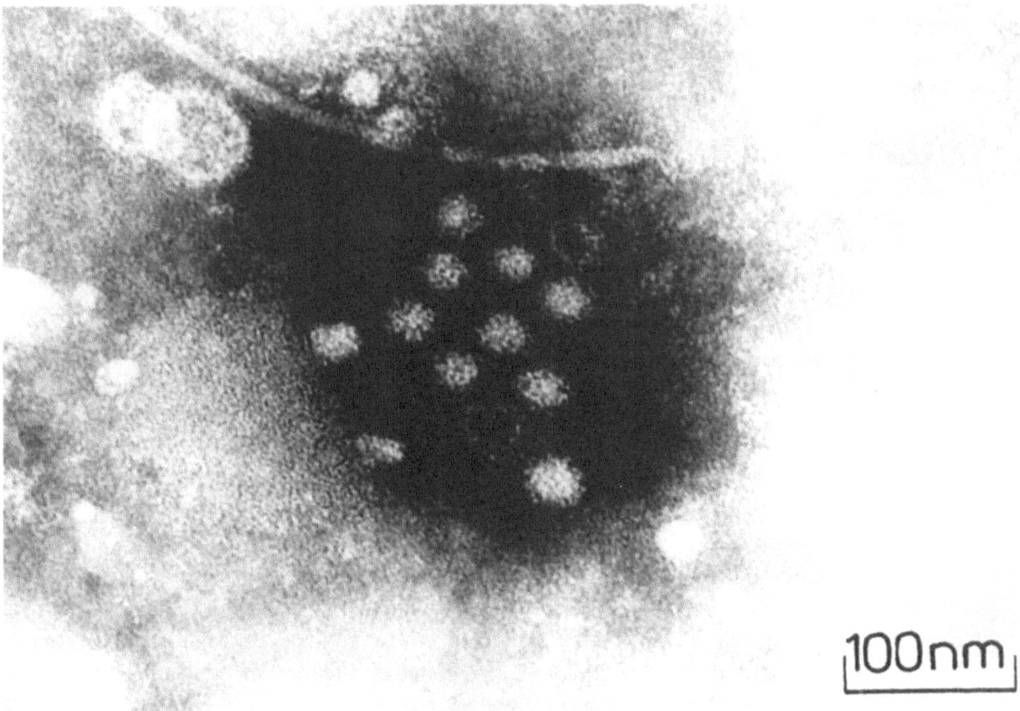

FIGURE 3. Norwalk virus free of antibody from a reference fecal sample. Note the amorphous surface structure and ragged edge to the particles. (Stained with 1.5% phosphotungstic acid pH 6.5.) (Courtesy of Dr. R. Dolin who supplied the reference fecal sample.)

Subsequent to the discovery and description of the Norwalk virus the American workers described two further, morphologically indistinguishable virus particles in stools collected from outbreaks of nonbacterial gastroenteritis which occurred in Hawaii and Montgomery County.[60] These particles were called Hawaii virus and Montgomery County virus, respectively, and produced symptoms indistinguishable from those produced by Norwalk virus. Serological studies established that Norwalk virus and Montgomery County virus were antigenically similar but distinct from Hawaii virus.[50,60] Thus at least two serotypes of the Norwalk virus group exist in America. Transmission of the Norwalk virus occurs by the fecal/oral route but the demonstration of virus in vomit by electron microscopy[61] offers an important alternative mode of spread which may explain the epidemic nature of these infections. Similar observations of Norwalk-like viruses in vomit have also been recorded in the U.K.,[61a] but in general detection of virus in such specimens is uncommon.[61b]

Initially, on the basis of certain biological characteristics the Norwalk virus was thought to be a parvovirus.[5,47,48] This tentative classification remained for some years and many review articles have compounded this classification. However on the basis of its morphology from an early micrograph, it was suggested that Norwalk virus was a calicivirus[10] and recent comparative studies with an animal calicivirus showed that Norwalk virus, although lacking the typical appearance of caliciviruses, closely resembled the appearance of this group of viruses.[7,62,63] Subsequent studies have confirmed this morphological classification and have drawn a clear distinction between Norwalk virus and known animal parvoviruses.[7] Recently more definitive evidence has shown that the virus possesses a single major protein with a molecular weight of 59,000[64] which is similar to the structural proteins of the Caliciviridae. It is important to confirm this finding but it may prove difficult in view of the small amounts of antigen excreted in stools. Further studies and definitive classification may have to await the propagation of Norwalk virus in cell culture to establish whether it contains an RNA genome. In the meantime, the habit of grouping Norwalk and Norwalk-like viruses with the small round featureless parvovirus-like particles should be discontinued.

The role of the Norwalk group of viruses in the etiology of acute gastroenteritis in children is negligible when compared with rotavirus but conversely the Norwalk group plays a major role in epidemic gastroenteritis of older children and adults in the community.[65] In contrast to most of the other candidate or proven enteric virus pathogens Norwalk and Norwalk-like viruses are usually excreted at such a low level that often electron microscopy is insufficiently sensitive for their consistent detection. Other than in America, laboratory workers rely entirely on electron microscopy for the detection of Norwalk-like viruses and this reliance on a relatively insensitive technique has been the most serious limitation to expanding the knowledge of the epidemiology of Norwalk and Norwalk-like viruses in epidemic gastroenteritis in other parts of the world. In all probability this group of viruses is grossly underdiagnosed.

Following the early descriptions of Norwalk, Montgomery County, and Hawaii viruses some years passed before other laboratory workers detected similar virus particles in human stools derived from outbreaks or sporadic cases of nonbacterial gastroenteritis.[7,62,63,66-70] It was occasionally established that Norwalk virus was the etiological agent in some of these outbreaks on the basis of antigenic studies carried out by immune electron microscopy.[70-71] However, the majority of outbreaks were associated with particles morphologically indistinguishable from Norwalk virus[7] and workers used the name Norwalk-like or small round structured viruses (SRSV) for these agents.[7,62,63,66-67] Considerable evidence has now accumulated that the Norwalk group of viruses are the causal agents of food-borne, water-borne, or shellfish-associated outbreaks of nonbacterial gastroenteritis.[62,65-67,69-75] In a large oyster-associated outbreak in Australia involving 2000 people, Murphy et al.[70] demonstrated that Norwalk virus was present in fecal samples by immune electron microscopy using paired sera from a chimpanzee experimentally infected with the Norwalk agent. However examination of the published micrographs show featureless particles with a smooth entire edge morphologically distinct from the obvious structured appearance of Norwalk virus. During the course of these electron microscopical studies the Australian workers noted the presence of a second, smaller virus measuring 22 to 25 nm in diameter in 50% of the stool samples. These particles resembled parvoviruses but were not considered to be significant by the Australian workers. This observation of a dual virus infection involving Norwalk and parvovirus-like particles has been reported by other workers.[7,66,74,75] Also, studies have confirmed that Norwalk-like as well as parvovirus-like particles can be incriminated in outbreaks associated with the consumption of shellfish.[74,75] In one report 82 to 85% of affected individuals involved in 14 outbreaks of shellfish-associated gastroenteritis were excreting either Norwalk-like or parvovirus-like viruses.[75]

Unfortunately it is unclear from this report how many of the 14 outbreaks were associated with Norwalk-like viruses as compared with the parvovirus-like particles. In view of the established etiological role of Norwalk and Norwalk-like viruses and the uncertainty of the role of the parvovirus-like agents in nonbacterial gastroenteritis this distinction is necessary for epidemiological purposes.

The laboratory detection of Norwalk and Norwalk-like viruses is notoriously difficult because of the low levels and short-term excretion of the virus in stools.[65] Similar observations have been made in British studies and the number of infected individuals who excrete Norwalk-like viruses in their stools can vary from as few as 10% to a maximum of 40% in any given outbreak.[61b] The low level of virus excretion, in conjunction with less than 50% of patients who may excrete virus detectable by electron microscopy in any given outbreak, suggests that these enteric pathogens are probably underdiagnosed.

Storage of stool samples containing Norwalk-like viruses is best undertaken at 4°C with an overlay of liquid paraffin[61b] since the morphological integrity of the particle deteriorates on freezing and thawing in contrast to the morphological stability of parvovirus-like particles under the same conditions. Techniques used in extracting Norwalk-like viruses from stool samples may also be important. In British studies either the double-spin ultracentrifugation technique[67] or ammonium sulfate precipitation[76] have been found to be the most sensitive techniques for the visualization of these viruses in the electron microscope. Although Norwalk-like viruses can be detected in unprotected cesium chloride gradients they have been found to be unstable after pelleting the virus from appropriate cesium chloride fractions.[77] This instability becomes even more important if stools contain Norwalk-like and parvovirus-like particles since it has been shown that the parvovirus-like particles are morphologically stable after pelleting from cesium chloride gradients.[77] The regular recovery of these viruses at the expense of the Norwalk-like viruses may lead to a misinterpretation of the true etiological role of the parvovirus-like particles in nonbacterial gastroenteritis.

D. Otofuke Agent

A 35- to 40-nm virus particle (Otofuke agent) with a visible substructure found in stool samples from an outbreak of gastroenteritis in an institution for mentally subnormal adults was reported by Japanese workers in 1979.[78] Significant sero-responses were detected in eight patients examined and the similarity of the viruses excreted in the stools of five patients was demonstrated by immune electron microscopy using acute and convalescent sera from one patient. The Otofuke agent was shown to be antigenically distinct from Norwalk virus, Wollan agent, and human calicivirus-like particles. The authors concluded that the Otofuke agent might represent a new candidate gastroenteritis virus. Subsequently these authors studied a further four outbreaks of acute gastroenteritis where virus particles similar in size and morphology to the Otufuke agent were detected by electron microscopy.[79] An antigenic relationship was demonstrated for all of the 35 to 40 nm particles found in these five outbreaks.

The description of these 35 to 40 nm virus particles is particularly interesting as the Japanese workers published micrographs from all the outbreaks showing the virus particles without any antibody added to their preparations. These viruses all had a feathery outline with a clear substructure and were morphologically indistinguishable from Norwalk and Hawaii virus which have also been examined without the addition of antibody.[7] The diameters of the viruses described by the Japanese workers are not incompatible with the diameter of the Norwalk and Hawaii agents, which have been estimated at 30 to 35 nm by British workers.[7,66,67] The buoyant density of these viruses (1.35 to 1.37 g/cm^3) is also in agreement with the density of Norwalk virus in cesium chloride. Although the Otofuke agent does not cross-react with the Norwalk virus, the possibility that the Otofuke agent is a new serotype of the Norwalk group may still be applicable.

E. Sapporo Agent

A further small round virus was reported by Japanese workers as a consequence of a study of nonbacterial gastroenteritis which occurred in an orphanage in Sapporo.[79] The Sapporo agent measured 33 to 39 nm in diameter and was recognized in stool samples from 17 of 35 patients examined. The virus possessed irregular surfaces made up of "fine radial projections"[79] and had a buoyant density of 1.37 to 1.40 g/cm^3 in cesium chloride. Evidence for a causal role was established by the demonstration of seroconversions in patients involved in this outbreak. An antigenic relationship to the Otofuke agent was also demonstrated by immune electron microscopy but a cross-reaction with Norwalk virus, Hawaii virus, Wollan virus, and a human calicivirus was not established.

Electron micrographs of the Sapporo agent free of antibody were published by the Japanese workers. The amorphous surface structure described by these workers was clearly visible and appears indistinguishable from that seen with Norwalk and Norwalk-like viruses.[7,66,67] The feathery outline of all the small round structured viruses that are morphologically similar to Norwalk virus makes accurate virus measurements difficult and may explain the small differences in recorded sizes.

F. Marin Agent

A 27-nm virus particle termed the Marin agent[80] was detected in the stools of hospitalized geriatric patients suffering from nonbacterial gastroenteritis. The outbreak occurred in a convalescent hospital located in Marin County in Northern California. Approximately 50% of the residents and a little under 20% of employees in this hospital were affected. The Marin agent was detected in the stools of 4 of 32 patients examined and serological responses were demonstrated in 16 of 18 patients tested by immune electron microscopy. Serological studies of a second large outbreak in a nearby county suggested that an antigenically similar agent to the Marin virus was involved. The authors concluded that the Marin agent was morphologically similar to but antigenically unrelated to Norwalk and Hawaii viruses.

It is unfortunate that these authors only published micrographs showing virus/antibody aggregates derived from their immune electron microscopy studies. Antibody coating of virus particles is known to interfere with the visualization of virus structure,[7,9] making it difficult to confirm the morphological similarity of the Marin agent to Norwalk virus. Some of the virus particles published by Oshiro and colleagues[80] of the Marin agent show some resemblance to astroviruses, and the publication of a micrograph showing the Marin agent free of antibody is awaited with interest. The single size estimation of 27 nm for the Marin agent is more compatible with the uniform size of astroviruses (28 $\pm$ 1.0 nm) rather than Norwalk virus.

G. Snow Mountain Agent

Immune electron microscopy was used for the detection of a 27-nm virus (Snow Mountain agent) in the stools of two out of five specimens examined from an outbreak of nonbacterial gastroenteritis in Colorado. The outbreak occurred in a resort camp in Snow Mountain and a high attack rate and secondary spread affecting over 400 people was noted.[81] Volunteer studies established a transmissible agent in a fecal filtrate derived from a primary case which produced illness in 75% of the volunteers. The 27-nm particle was also demonstrated in the stools of some of these ill volunteers and seroconversions to the Snow Mountain agent were demonstrated in sera collected from both the primary cases as well as from the volunteers. A serological relationship to Norwalk and Hawaii virus could not be demonstrated by immune electron microscopy. The authors concluded that the Snow Mountain agent was morphologically similar but antigenically different from Norwalk and Hawaii viruses.

It is difficult to comment on the Snow Mountain agent described by these workers as only micrographs showing the agent aggregated by antibody were published. These studies, in common with those carried out on the Marin agent, highlight the importance of examining small round viruses in the electron microscope before the addition of antibody. This will ensure that structural details on the virus will be revealed and will allow a correct classification.

III. SUMMARY AND DISCUSSION OF STRUCTURED VIRUSES

The division of small round viruses into two major groups (structured and featureless) in the interim classification scheme proposed by Caul and Appleton[7] has resolved some of the anomalies which have occurred in the literature. Within the structured viruses described above, a number of viruses can be confidently and definitively identified.

The identification of astroviruses relies entirely on the demonstration of the characteristic surface star. High contrast micrographs are essential as they greatly enhance the appearance of this surface structure. It is intriguing that astroviruses have been recognized and identified by many electron microscopists in the U.K. and elsewhere in the world but not by American workers. This can probably be explained by the fact that immune electron microscopy, which masks any surface structure, is commonly carried out in the U.S. The smooth entire edge possessed by astroviruses allows accurate virus measurements and explains the narrow size range of 27 to 29 nm reported in the literature.

Similarly, caliciviruses present no problems in identification when they exhibit their classical appearance of surface stain-filled cups. It is noteworthy that within a population of calicivirus particles many will not show a classical appearance and often show an amorphous structure without recognizable stain-filled cups. This holds true even when pure, cell-culture propagated, caliciviruses of feline origin are examined in the electron microscope.[62,63] In view of this gross variation in morphology of the caliciviruses it is unnecessary to postulate that more than one virus is present in a fecal sample if the electron microscopist is familiar with the variation that can occur within a population of purified caliciviruses. It is in this context that comparative studies using known virus reference strains are so important. The ragged outline of the caliciviruses makes accurate virus particle measurements extremely difficult and it is not surprising that different workers will calculate different sizes with a wide range of diameters for the same virus. In addition electron microscopes can vary from 5 to 20% in their indicated magnification settings which makes instrument-to-instrument variation significant when comparisons of virus particle diameters are attempted. If such comparisons are carried out then different viruses should be examined in the same electron microscope using standard calibration specimens. Only in this way will meaningful differences in size become apparent.

The amorphous reproducible surface structure of the Norwalk group of viruses does not allow a definitive morphological classification. In spite of this situation, this group of viruses closely resemble known reference strains of caliciviruses lacking the classical appearance of stain-filled cups. The American workers estimated a particle diameter of approximately 27 nm for the Norwalk group but British workers suggest that they measure 30 to 35 nm in diameter.[7,66,67] This apparent difference in size can be explained by the ragged outline of the Norwalk group of viruses which does not allow for accurate measurements, as well as instrument variation and preparative techniques. In light of the problems associated with accurate measurements of virus particles with a ragged outline, emphasis on size differences between small round structured viruses should be interpreted with extreme caution. It is unacceptable at present that the small differences in size should be the major reason for the Sapporo and Otofuke agent to be classified as different from the Norwalk group of viruses. Although they are antigenically distinct from Norwalk virus, it is still feasible that they represent new serotypes. Morphologically they appear indistinguishable from the Norwalk group of viruses. In the U.K. small round virus particles which are morphologically indistinguishable from Norwalk virus are called small round structured viruses (SRSV) (Figure 4) merely for convenience and reproducibility in reporting for epidemiological purposes. Other workers elsewhere in the world describe such viruses as Norwalk-like. Both terms are interchangeable and describe the same virus/viruses. It is no longer acceptable to classify the Norwalk group of viruses as parvoviruses or to persistently group them with the "featureless" viruses which are more likely, within our present knowledge, to be correctly classified as possible parvoviruses.

It is inappropriate at the present time to discuss either the Marin County or Snow Mountain agent, since micrographs have not been published demonstrating the morphology of these two agents free of antibody coating. It has been previously stated that surface structure is masked by antibody and the importance of examining small round viruses in the absence of antibody can only be reemphasized here.

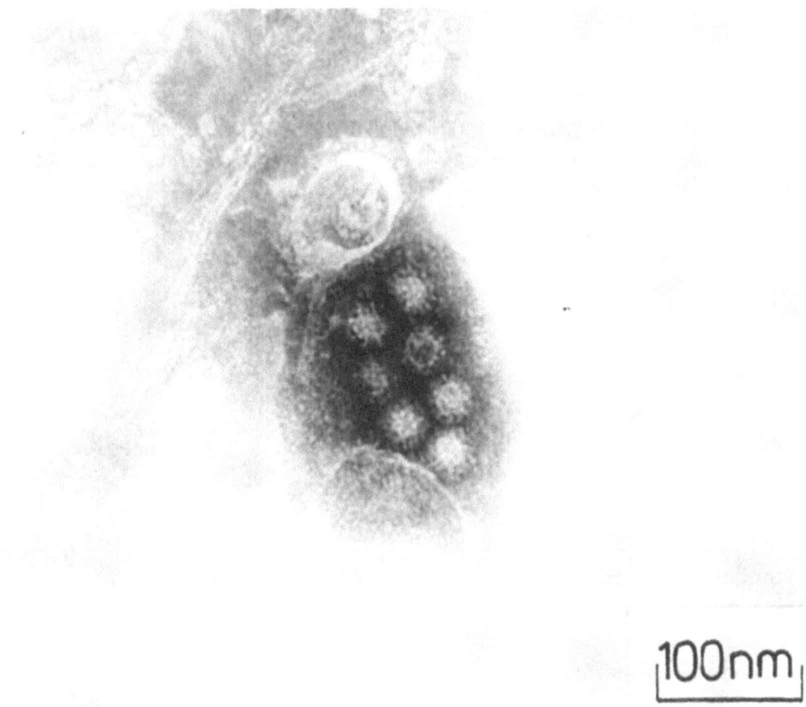

FIGURE 4. Typical particles, free of antibody, morphologically indistinguishable from Norwalk virus found in an outbreak of nonbacterial gastroenteritis. These particles are reported as Norwalk-like or "small round structured viruses" (SRSV) in the U.K. Prepared by ammonium sulfate precipitation. (Stained with 1.5% phosphotungstic acid pH 6.5.)

Considerable data has now accumulated on the etiological association with gastroenteritis of all the structured viruses described to date. The Norwalk and Norwalk-like viruses are clearly the most important pathogens with respect to clinical disease and give rise to large food-borne and community-wide outbreaks of gastroenteritis. The extensive volunteer and community studies carried out by the American workers have clearly established the Norwalk group of viruses as enteric pathogens and the reader is referred to the comprehensive data already published on this group of viruses.[65] Studies elsewhere in the world have confirmed the original observations of the American group and morphologically indistinguishable Norwalk-like virus particles have been recorded by many workers in outbreaks of gastroenteritis. It is unclear at present how many serotypes of the Norwalk group exist, but current indications are that there are more than the two which have been described by the American workers. The recent identification of a 59,000 mol wt protein in preparations of Norwalk virus support its proposed classification as a calicivirus. One of the most exciting areas of research in the future will be to establish the complete characterization of the Norwalk group of viruses and confirm or refute their classification as caliciviruses. Astroviruses and caliciviruses have been shown to have a causal role in gastorenteritis. Volunteer studies in adults have confirmed that astroviruses are pathogenic but they appear to have a lower pathogenicity for adults than for children. The high incidence of antibody in the community and the relatively uncommon detection of astroviruses in fecal samples suggest that asymptomatic infection is more common. Similar observations apply to caliciviruses which are detected in the electron microscope even less often than astroviruses. Asymptomatic infection is probably quite common with this group of viruses as well.

The Otofuke and Sapporo agents reported by the Japanese workers appear antigenically similar and are likely to be the same virus. Their etiological role is strongly supported by the epidemiological data reported by these workers. It remains to be clarified whether these two agents are distinct from the Norwalk group of viruses as suggested by the Japanese workers rather than a different serotype. Further data is required on the morphology of both the Marin County and Snow Mountain agents before relationships to the other structured viruses can be discussed. In conclusion, considerable data has been accumulated on the structured viruses but many of their fundamental properties remain to be elucidated before a definitive classification can be achieved. A summary of the properties of the "structured" viruses is shown in Table 3.

IV. FEATURELESS VIRUSES

A. Wollan Agent

The first report of "featureless" small round viruses in human feces relates to the studies carried out on an outbreak of winter vomiting disease which occurred in a boy's boarding school in England. Fecal filtrates from this outbreak were administered to volunteers and a transmissible agent was demonstrated.[6] The name Wollan was assigned to this (Figure 5) which produced a clinical picture indistinguishable from that seen in volunteers infected with the Norwalk agent. Examination of fecal samples from volunteers infected with the Wollan agent revealed a hitherto undescribed small round parvovirus-like particle measuring approximately 22 nm in diameter.[6,82] This report closely followed the American studies on Norwalk virus[5] which was also described as parvovirus-like, but it rapidly became clear that both these viruses could not be parvoviruses as marked differences in both size and more importantly in morphology existed. This situation became resolved over the ensuing years and Norwalk virus is now considered to be a candidate calicivirus rather than a parvovirus. Further studies with the Wollan agent demonstrated that it banded at a density of 1.38 g/cm^3 and in comparative studies with known animal parvoviruses no differences in morphology could be found.[82] Consideration was given at this time to the possibility that the Wollan agent may represent a member of the tail-less RNA or DNA bacteriophages. Consequently representatives of these small RNA and DNA bacteriophages (ØX 174 and MS2) were examined in the electron microscope and were shown to be morphologically distinct from the Wollan agent.[82] Further evidence to support these morphological studies came from immune electron microscopy. Human sera used to aggregate the Wollan agent did not react with these two representative RNA and DNA bacteriophages. The possibility that a currently unrecognized bacteriophage exists which has the same physical and morphological properties as the Wollan agent cannot at present be excluded. The scarcity of virus particles in stool samples and the lack of a suitable in vitro system for the propagation of the Wollan agent has hampered any biochemical characterization of this virus.

Sero-epidemiological studies with the 22-nm human fecal parvovirus-like particles showed that 36% of children aged 10 to 15 years possessed the antibody.[83] At this time, a new human serum parvovirus-like particle was described in asymptomatic blood donors[84-85] and studies with these human serum parvovirus-like particles and the fecal virus showed that they were antigenically distinct[83] although identical in size and morphology. In addition their similar buoyant densities in cesium chloride as well as their similarities in comparative electron microscopical studies with animal parvoviruses led the British workers to conclude that the fecal virus might be a new human parvovirus.[82] Recent studies using monoclonal antibody raised against the human serum parvovirus has not revealed any cross-reactivity with the human fecal virus.[61b]

It is clear from the volunteer studies carried out with the Wollan agent[86] that a transmissible agent can be demonstrated. Clinically the illness was indistinguishable from that produced

Table 3
PROPERTIES OF SMALL ROUND STRUCTURED VIRUSES

Virus	Surface structure	Particle size	Buoyant density in cesium salts	Nucleic acid	Classification
Astrovirus	5/6 Pointed surface star	28 ± 1.0 nm	1.39—1.40	RNA[a]	New group intermediate between Picornavirus and Calicivirus
Calicivirus	"Star of David" stain-filled cups	31—38 nm	1.36—1.40	RNA	Calicivirus
Norwalk, Hawaii, Montgomery County	Amorphous, no geometric symmetry	30—35 nm	1.36—1.41	Presumed[b] RNA	Possible Calicivirus
Otofuke agent	Amorphous	35—40 nm	1.36—1.37	?	Possible Calicivirus
Sapporo agent	Amorphous	33—39 nm	1.37—1.40	?	Possible Calicivirus
Snow Mountain[c] agent	Not defined	27 nm	?	?	?
Marin agent[c]	Not defined	27 nm	?	?RNA	?Astrovirus

[a] Identified from lamb astrovirus particles.
[b] RNA presumed from data on protein composition indicating a 59,000 mol wt protein.
[c] Authors describe agents as "Norwalk"-like but morphology unresolvable because of antibody coating.

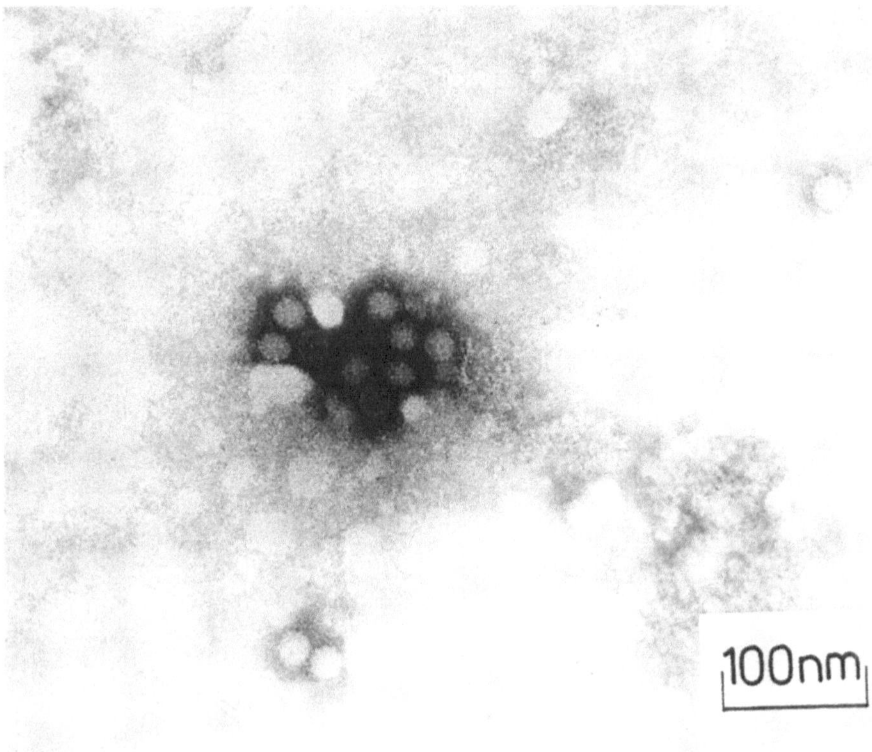

FIGURE 5. "Featureless" parvovirus-like particles from feces. Note the marked difference in morphology when compared with Norwalk and the other "structured" viruses. (Stained with 1.5% phosphotungstic acid pH 6.5.)

by Norwalk virus but the detection of a 22-nm parvovirus-like particle in fecal samples from symptomatic volunteers does not constitute proof of an etiological role for that virus. Further studies are required, using purified virus preparations for administration to volunteers, similar to those carried out on the Norwalk virus[65] to strengthen what at this stage is only an apparent association with gastroenteritis. Indeed, the British workers were unable to find an association between the 22-nm parvovirus-like particle and gastroenteritis since indistinguishable virus particles were observed in the stools of healthy individuals.[6] It has been suggested that 9% of normal individuals may excrete these 22-nm parvovirus-like particles.[75] This asymptomatic excretion does not exclude a causal role in human gastroenteritis as asymptomatic excretion of canine parvoviruses[87] is also documented where an etiological role in gastroenteritis has been clearly established.[88]

B. Ditchling Agent

Following the description of the Wollan agent, Appleton et al.[59] described the results of their studies on an outbreak of winter vomiting disease which occurred in a primary school in the village of Ditchling in southeast England. Fecal samples from this outbreak were examined by electron microscopy and small round "featureless" particles measuring approximately 26 nm in diameter (Ditchling agent) were observed. The detection of these particles was facilitated by their aggregation following the addition of human serum to cesium chloride purified virus.[59] The resemblance these particles had to the picornavirus group was noted but this classification was excluded on the basis of a buoyant density of 1.38 to 1.40 g/cm^3 in cesium chloride. The possibility that these particles may have been bacteriophages was also discussed by these workers. Serological studies using human sera derived from the Ditchling outbreak and from the Wollan volunteer studies showed that the Wollan and Ditchling agents were antigenically similar. Immune electron microscopy using paired sera from volunteers in the Norwalk and Hawaii studies as well as paired sera from a chimpanzee experimentally inoculated with the Norwalk agent showed no cross-reactivity with the Ditchling agent. Electron micrographs of the Ditchling agent free of antibody were not reproduced in this publication but subsequently it was grouped as a small round "featureless" virus morphologically distinct from both Norwalk and Hawaii viruses.[7] Recent comparative studies have shown that the Ditchling agent is indistinguishable from the Wollan agent and known parvoviruses in size, buoyant density, and morphology,[7] but distinct from Norwalk and Hawaii viruses.

The continual excretion of the Ditchling agent in two of the children 1 month after their symptoms had ceased is also in agreement with the observations of the British workers with the Wollan agent, but in contradiction to the excretion pattern of Norwalk virus which is short-lived.[65] In view of the documented persistent excretion of the Ditchling agent in human feces it is as difficult to assign a causal role in human gastroenteritis for this agent as it is for the Wollan agent. In support of these conclusions it is noteworthy that rising titers have only very rarely been demonstrated in paired sera collected from either the Wollan or Ditchling outbreaks. This is also at variance with the serological studies carried out with Norwalk virus where increases in antibody levels have been demonstrated in paired serum samples from infected individuals.[65]

C. Cockle Agent

Outbreaks of food-borne nonbacterial gastroenteritis associated with the consumption of shellfish have been reported more frequently in recent years. The first outbreaks to be studied using electron microscopy were associated with the consumption of cockles in the south of England.[73] Fecal samples from three distinct outbreaks were studied and small round "featureless" virus-like particles (Cockle agent) measuring 25 to 26 nm in diameter with a buoyant density of 1.40 g/cm^3 in cesium chloride were detected in almost all the samples examined. Subsequently, comparative studies have established that the Cockle agent is indistinguishable from the Wollan and Ditchling agents as well as from known parvoviruses in size, morphology, and buoyant density.[7] These same studies demonstrated the marked differences in size and morphology which exist between these featureless parvovirus-like particles and the Norwalk group of viruses. Serological studies by immune electron microscopy suggested that the Cockle agent is antigenically distinct from both the Wollan and Ditchling agents.[73] Persistent excretion of the Cockle agent was not documented by Appleton and Pereira in these cockle-associated outbreaks of gastroenteritis, but establishment of a causal role for this agent requires further studies similar to those discussed for the Wollan and Ditchling agents. The likelihood that shellfish will concentrate more than one small round virus from their polluted environment is a real possibility.

D. Parramatta Agent

Christopher et al.[89] studied an outbreak of nonbacterial gastroenteritis in a primary school in Australia where approximately 50% of the pupils and staff became ill. Small round viruses measuring 23 to 26 nm in diameter (Parramatta agent) were detected in 61% of the stools examined from this outbreak. The authors described the particles as Norwalk-like and demonstrated antibody rises in acute and convalescent sera to the virus by immune electron microscopy. Further serological studies did not establish any antigenic relationship to Norwalk virus. It was noted by these workers that excretion of the Parramatta agent continued for some weeks after cessation of symptoms — a feature common to the Wollan and Ditchling agents but in sharp contrast to Norwalk virus. The authors concluded that the Parramatta agent was a human parvovirus which represented a new disease entity in Australia. Recent

Table 4
PROPERTIES OF SMALL ROUND "FEATURELESS" VIRUSES (CANDIDATE PARVOVIRUSES) AND REFERENCE STRAINS OF PARVOVIRUSES AND PICORNAVIRUSES

Virus	Surface structure	Size range	Buoyant density in cesium salts	Nucleic acid
Wollan agent	Smooth	20—22 nm	1.38	?
Ditchling agent	Smooth	26 nm	1.38—1.40	?
Cockle agent	Smooth	25—26 nm	1.40	?
Paramatta agent	Smooth	23—26 nm	?	?
Mink enteritis virus (reference parvovirus)	Smooth	18—26 nm	1.38—1.48	DNA
Serum parvovirus	Smooth	22—25 nm	1.36—1.40	DNA
Picornavirus (coxsackie B5)	Smooth	20—30 nm	1.34	RNA

comparative studies have shown that the Parramatta agent lacks any surface structure, appears "featureless" in the electron microscope, and is indistinguishable from Wollan, Ditchling, and the Cockle agents but distinct from Norwalk virus in both size and morphology.[7] From these comparative studies the Parramatta agent has the morphology, size, and buoyant density consistent with known parvoviruses.

V. SUMMARY AND DISCUSSION OF "FEATURELESS" VIRUSES

It has become clear over recent years that considerable confusion exists in the interrelationships of many of the small round viruses and some have been grouped together on the basis of particle size, buoyant density, and similarity of the clinical illness. The most outstanding example, is the grouping of Norwalk and Hawaii virus (calicivirus candidates) with the morphologically distinct Ditchling, Wollan, and Parrammata agents. In addition, significant serological responses have been demonstrated for the Norwalk group of viruses but not for most of the described parvovirus-like particles. Measurement of virus particles is an unreliable marker for grouping viruses, as distinct virus families may have similar particle diameters, e.g., parvoviruses and picornaviruses. Similarly, buoyant density estimations, although useful for distinguishing between potential parvoviruses and picornaviruses, can be misleading as taxonomically distinct virus groups exhibit similar values, e.g., parvoviruses and caliciviruses. The interim scheme for subgrouping the various small round viruses[7] has in practice been useful for distinguishing the variety of agents which have been described. The "structured" viruses (Table 3) are gradually becoming characterized as probable RNA viruses and we are left with a "featureless" group of viruses which remain the only realistic group which can be considered as possible parvoviruses.

Far less data is available for firm conclusions to be drawn on the candidate parvovirus group of human fecal viruses (Table 4). Certainly the "featureless" morphology of this group allows ready differentiation from all of the "structured" viruses. Furthermore, comparisons with known animal parvoviruses are compatible with the morphology of these human candidate fecal parvoviruses. The buoyant density of Wollan, Ditchling, Parramatta, and Cockle agents in conjunction with their size and common morphology (Figure 5) tend to support the suggestion that they may indeed be parvoviruses.[7] It is clear that all of these latter candidate parvoviruses differ considerably from both Norwalk and Norwalk-like viruses in both size but more importantly in morphology.[7] The similar buoyant densities of the two different groups serves as a warning that this property can be misleading when used as a marker of relationships.

These comparative studies do not however allow a definitive classification at this stage. Considerable data on protein composition and genome analysis is required before these candidate human fecal parvoviruses can be confidently assigned to their appropriate virus family. The lack of a suitable in vitro system means that this biochemical data will not be readily forthcoming and the very low levels of virus excretion in stools reinforces the urgency for a suitable in vitro system to be developed. It is disappointing that no antigenic similarities with the human serum parvovirus have yet been revealed either by immune electron microscopical studies on the fecal and serum viruses[83] or by means of monoclonal antibodies raised against the human serum parvovirus.[61b] These studies do not exclude the final classification of these enteric viruses as parvoviruses as other parvovirus serotypes may exist. Studies on two common RNA and DNA bacteriophages (ØX 174 and MS 2) support opinions that these human candidate parvoviruses are not members at least of these two representative groups of bacteriophages. Although unlikely, the possibility remains that they may represent members of hitherto unknown tail-less bacteriophages.

In contrast to the Norwalk virus, conclusive evidence that these human enteric candidate parvoviruses cause gastroenteritis is lacking. The frequent observation that persistent excretion of these viruses occurs in healthy individuals[3,6] and for weeks after cessation of symptoms[59,89] makes the assignment of a causal role difficult. Volunteer studies have not been conducted on the Ditchling, Parramatta, and Cockle agent, so proof of transmissibility with associated symptoms is unavailable at present. Although the Wollan agent has been studied in volunteers, and clinical illness indistinguishable from that produced by Norwalk virus demonstrated, it remains unclear that the candidate human fecal parvovirus (Wollan agent) visualized in the electron microscope was the etiological agent. Further studies using purified Wollan virus similar to that carried out with the Norwalk virus is clearly indicated before a causal role can be assigned to the human fecal candidate parvovirus. In the past, Wollan virus has been considered to be a member of the "Norwalk" group of viruses and by analogy with the comprehensive studies carried out by Kapikian and colleagues[65] on the Norwalk virus it has been assumed that Wollan virus was an enteric pathogen. The considerable differences which are now apparent between these two different viruses highlight the need for a reappraisal of the role of Wollan and the other human fecal candidate parvoviruses such as Ditchling, Cockle, and Parramatta agents in their role in nonbacterial gastroenteritis. Furthermore, serological evidence of recent infection with the Wollan, Ditchling, and Cockle agents is also lacking, in sharp contrast to the epidemiological data available on Norwalk virus infections where consistent seroconversions have been demonstrated. Although an etiological role for the human fecal candidate parvoviruses has not been established, studies in animals have conclusively demonstrated that animal parvoviruses cause gastroenteritis. Replication of the canine parvovirus in small intestinal crypt cells gives rise to a severe villous atrophy with a resultant high mortality in the neonatal animal.[88] Chronic excretion of the canine parvovirus in asymptomatic dogs is well documented[87] and in view of this data, chronic excretion in asymptomatic human patients is not a good reason for excluding a causal role for the human fecal candidate parvovirus. It does, however, make it difficult to prove, as a suitable model in man similar to those used by the veterinary profession for studies on the pathogenesis of animal parvoviruses is not available. It is also intriguing that the highest incidence of morbidity and mortality with the animal enteric parvoviruses occur in the neonatal animal. An analogous situation in man has not yet been described, as all the descriptions of human infection relate to schoolchildren or adults. The most extensively studied human fecal candidate parvoviruses are the Wollan, Ditchling, Cockle, and Parramatta agents. Many other workers have described "small round viruses" in stools of neonates or young children in sporadic cases of gastroenteritis. When these are seen in association

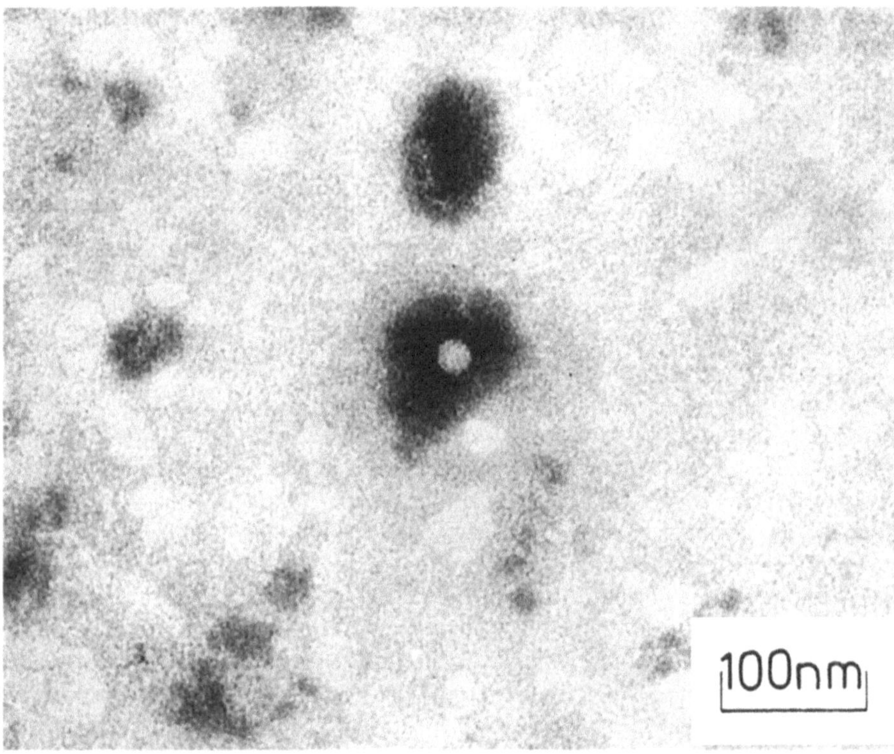

FIGURE 6. "Featureless" parvovirus-like particle also present in reference stool containing Norwalk virus (see Figure 3). Large numbers were detected in this sample (see Reference 7). (Stained with 1.5% phosphotungstic acid pH 6.5.)

with adenoviruses it is likely, but not proof, that these small round viruses are adeno-associated viruses which are known members of the Parvoviridae. Enteric adenoviruses are now recognized as causal agents of childhood gastroenteritis but have rarely been described in adults. It is unlikely, but unproven, that the human fecal candidate parvoviruses seen in stools of adults are adeno-associated viruses.

The role of the human parvovirus-like particles in nonbacterial gastroenteritis is further obscured by the observation that these viruses are commonly present in fecal samples derived from outbreaks in which either Norwalk or Norwalk-like viruses are incriminated. In an outbreak of gastroenteritis involving two hospitals in the south of England, a Norwalk-like virus was identified and was considered to be the cause of the outbreak.[66] The Norwalk-like virus was only identified in symptomatic patients but a smaller "featureless" parvovirus-like particle was also detected in stools from both symptomatic and asymptomatic patients who were studied as a control group. Similar observations were noted by Murphy et al.[70] during the course of studies carried out on a large oyster-associated outbreak of gastroenteritis in Australia. In a further study of an oyster-associated outbreak of viral gastroenteritis in Australia, Linco and colleagues[74] described Norwalk virus and a smaller "featureless" parvovirus-like particle in the same stool sample. It is of interest that these workers noted not only the difference in size but also the morphological differences of these two different viruses. In a recent comparative study of small round viruses,[7] reference stools of Norwalk virus (Figure 3) and Hawaii virus were also shown to contain smaller "featureless" parvovirus-like particles which are clearly morphologically distinct (Figure 6) from Norwalk and Hawaii viruses. Studies of many outbreaks of nonbacterial gastroenteritis associated with Norwalk-like particles in Manchester have also shown that small round "featureless" parvovirus-like particles can be detected in the stools of patients with or without symptoms[61a] and similar observations have been noted in our own studies of Norwalk-like outbreaks of nonbacterial gastroenteritis.[74a]

It is clear that many outbreaks, where Norwalk or Norwalk-like viruses can be detected in stool samples, may also contain morphologically distinct small round parvovirus-like particles in the stools of both symptomatic and asymptomatic individuals. It is an interesting exercise to speculate on the association, if any, of these two morphologically distinct viruses. Enteric animal parvoviruses have been shown to infect and replicate in crypt cells of the small intestine and the resultant damage to these crypt cells can give rise to a severe villous atrophy. Parvoviruses are known to require rapidly dividing cells for their replication and crypt cells provide the ideal cell environment for virus multiplication. The frequent observation of human parvovirus-like particles both in asymptomatic and symptomatic patients weeks after symptoms have ceased suggests a chronic carrier state. Assuming that a low-level parvovirus infection of crypt cells is occurring, then any viral damage to the villous epithelium would result in a crypt cell hyperplasia — a cellular state which is ideal for the replication of any parvoviruses present. Replication of Norwalk virus is thought to occur in the villous epithelium giving rise to a villous atrophy and a crypt cell hyperplasia,[65] and this group of enteric pathogens could provide the mechanism which would allow for the replication of large numbers of parvoviruses. This hypothesis would be one explanation for the frequently observed association of the human fecal candidate parvovirus particles and Norwalk or Norwalk-like viruses. It has also been suggested that these parvovirus-like particles represent degraded Norwalk viruses. This however would not explain the presence of these parvovirus-like particles in the stools of asymptomatic adults for many weeks after cessation of symptoms. Furthermore, serological studies do not support this suggestion where the human parvovirus-like particles have been shown to be antigenically distinct from Norwalk virus. This commonly observed dual virus infection raises severe problems in the development of antigen/antibody assays which are based on human sera and unpurified antigens in stool samples. Serological studies have shown that at least two serotypes of the human parvovirus-like particles exist.[73] It remains to be established whether other serotypes can be added to this list. Future studies on the biology and biochemistry of these viruses will be hampered by the scarcity of virus particles in stools and the lack of a suitable in vitro system for amplification of the virus. The complete dependency on the relatively insensitive technique of electron microscopy for the study of these parvovirus-like particles highlights the need for alternative assay systems which would increase our knowledge on the epidemiology of these poorly understood agents. It may be that the most informative approach in the future for the study of these candidate human fecal parvoviruses will be either the application of monoclonal antibodies derived from the human serum parvovirus or DNA hybridization with known parvoviruses.

REFERENCES

1. **Tyrrel, D. A. J., Bynoe, M. L., and Hoorn, B.,** Cultivation of "difficult" viruses from patients with common colds, *Br. Med. J.*, 1, 606, 1968.
2. **Egglestone, S. I.,** Laboratory Investigation Into the Aetiology of Non-bacterial Gastroenteritis, Ph.D. thesis, University of Bristol, England, 1968.
3. **Paver, W. K.,** An Investigation Into the Cause of Nonbacterial Gastroenteritis by the Techniques of Immunofluorescence and Immune Electron Microscopy, Ph.D. thesis, University of Bristol, England, 1975.
4. **Rubenstein, D. B. and Tyrrell, D. A. J.,** Growth of viruses in organ culture of intestines, *Br. J. Exp. Pathol.*, 51, 210, 1978.

5. **Kapikian, A. Z., Wyatt, R. G., Dolin, R., Thornhill, T. S., Kalica, A. R., and Chanock, R. M.,** Visualization by immune electron microscopy of a 27-nm particle associated with acute infectious nonbacterial gastroenteritis, *J. Virol.*, 10, 1075, 1972.
6. **Paver, W. K., Caul, E. O., Ashley, C. R., and Clarke, S. K. R.,** A small virus in human faeces, *Lancet*, 1, 237, 1973.
7. **Caul, E. O. and Appleton, H.,** The electron microscopical and physical characteristics of small round human fecal viruses: an interim scheme for classification, *J. Med. Virol.*, 9, 257, 1982.
8. **Madeley, C. R. and Cosgrove, B. P.,** 28 nm particles in feces in infantile gastroenteritis, *Lancet*, 2, 451, 1975.
9. **Ashley, C. R., Caul G. O., and Paver, W. K.,** Astrovirus associated gastroenteritis in children, *J. Clin. Pathol.*, 31, 939, 1978.
10. **Holmes, I. H.,** Viral Gastroenteritis, *Prog. Med. Virol.*, 25, 1, 1979.
11. **Middleton, P. J., Szymanski, M. T., and Petric, M.,** Viruses associated with acute gastroenteritis in young children, *Am. J. Dis. Child.*, 131, 733, 1977.
12. **McLean, D. M., Wong, K. S. K., and Bergman, S. K. A.,** Virions associated with acute gastroenteritis in Vancouver in 1976, *Can. Med. Assoc. J.*, 117, 1035, 1977.
13. **Madeley, C. R.,** Comparison of the features of astroviruses and caliciviruses seen in samples of feces by electron microscopy, *J. Infect. Dis.*, 139, 519, 1979.
14. **Ashley, C. R. and Caul, E. O.,** Potassium tartrate-glycerol as a density gradient substrate for separation of small round viruses from human feces, *J. Clin. Microbiol.*, 16, 377, 1982.
15. **Herring, A. J., Gray, E. W., and Snodgrass, D. R.,** Purification and characterisation of ovine astrovirus, *J. Gen. Virol.*, 53, 47, 1981.
16. **Flewett, T. H.,** Diagnosis of enteritis virus, *Proc. R. Soc. Med.*, 69, 693, 1976.
17. **Kjeldsberg, E.,** Small spherical viruses in faeces from gastroenteritis patients, *Acta Pathol. Microbiol. Scand. Sect. B*, 85, 351, 1977.
18. **Madeley, C. R. and Cosgrove, B. P.,** Viruses in infantile gastroenteritis, *Lancet*, 2, 124, 1975.
19. **Schnagl, R. D., Holmes, I. H., and Mackay-Scollay, E. M.,** A survey of rotavirus associated with gastroenteritis in aboriginal children in Western Australia, *Med. J. Aust.*, 1, 304, 1978.
20. **Kurtz, J. B., Lee, T. W., and Pickering, D.,** Astrovirus associated gastroenteritis in a children's ward, *J. Clin. Pathol.*, 30, 948, 1977.
21. **Konno, T., Suzuki, H., Ishida, N., Chiba, R., Mochizuki, K., and Tsunoda, A.,** Astrovirus-associated epidemic gastroenteritis in Japan, *J. Med. Virol.*, 9, 11, 1982.
22. **Maass, G., Baumeister, H. G., Heigeinoller, R., Jansen, P., and Schumacher, H.,** Visualization by electron microscopy of 28 nm virus particles (Astrovirus) in faeces of newborns with acute nonbacterial gastroenteritis, *Zentralbl. Bakteriol. Parasitenkd. Infektionskr, Hyg. Ab. Orig Reihe A*, 242, 423, 1978.
23. **Kurtz, J. B., Lee, T. W., Craig, J. S., and Reed, S. E.,** Astrovirus infection in volunteers, *J. Med. Virol.*, 3, 221, 1979.
24. **Snodgrass, D. R., Angus, K. W., Gray, E. W., Menzies, J. D., and Paul, G.,** Pathogenesis of diarrhoea caused by astrovirus infections in lambs, *Arch. Virol.*, 60, 217, 1979.
25. **Kurtz, J. and Lee, T.,** Astrovirus gastroenteritis age distribution of antibody, *Med. Microbiol. Immunol.*, 166, 227, 1978.
26. **Lee, T. W. and Kurtz, J. B.,** Astroviruses detected by immunofluorescence, *Lancet*, 2, 406, 1977.
27. **Lee, T. W. and Kurtz, J. B.,** Serial propagation of astrovirus in tissue culture with the aid of trypsin, *J. Gen. Virol.*, 57, 421, 1981.
28. **Kurtz, J. B. and Lee, T. W.,** Human astrovirus serotypes, *Lancet*, 2, 1405, 1984.
29. **Schaffer, F. L.,** Caliciviruses, *Comp. Virol.*, 14, 249, 1979.
30. **Suzuki, H., Konno, T., Kutsuzawa, T., Imai, A., Tazawa, F., Ishida, N., Katsushima, N., and Sakamoto, M.,** The occurrence of caliciviruses in infants with acute gastroenteritis, *J. Med. Virol.*, 4, 321, 1979.
31. **Madeley, C. R. and Cosgrove, B. P.,** Caliciviruses in man, *Lancet*, 1, 199, 1976.
32. **Flewett, T. H. and Davies, H.,** Calicivirus in man, *Lancet*, 1, 311, 1976.
33. **McSwiggan, D. A., Cubitt, W. D., and Moore, W.,** Caliciviruses associated with winter vomiting disease, *Lancet*, 1, 1215, 1978.
34. **Cubitt, W. D., McSwiggan, D. A., and Moore, W.,** Winter vomiting disease caused by calicivirus, *J. Clin. Pathol.*, 32, 786, 1979.
35. **Chiba, S., Sakuma, Y., Kogasaka, R., Akihara, M., Horino, K., Nakao, T., and Fukui, S.,** An outbreak of gastroenteritis associated with calicivirus in an infant home, *J. Med. Virol.*, 4, 249, 1979.
36. **Chiba, S., Sakuma, Y., Kogasaka, R., Akihara, M., Terashima, H., Horino, K., and Nakao, T.,** Fecal shedding of virus in relation to the days of illness in infantile gastroenteritis due to calcivirus, *J. Infect. Dis.*, 142, 247, 1980.
37. **Oishi, I., Maeda, A., Yamazaki, K., Minekawa, Y., Nishimur, H., and Kituara, T.,** Calicivirus detected in outbreaks of acute gastroenteritis in school-children, *Biken J.*, 23, 163, 1980.

38. **Cubitt, W. D., McSwiggan, D. A., and Arstall, S.,** An outbreak of calicivirus infection in a mother and baby unit, *J. Clin. Pathol.*, 33, 1095, 1980.
39. **Spratt, H. C., Marks, M. I., Gomersall, M., Gill, P., and Pai, C. H.,** Nosocomial infantile gastroenteritis associated with mini rotavirus and calicivirus, *J. Pediatr.*, 93, 922, 1978.
40. **Cubitt, W. D., Pead, P. J., and Saeed, A. A.,** A new serotype of calicivirus associated with an outbreak of gastroenteritis in a residential home for the elderly, *J. Clin. Pathol.*, 34, 924, 1981.
41. **Sakuma, Y., Chiba, S., Kogasaka, R., Terashima, H., Nakamura, S., Horino, K., and Nakao, T.,** Prevalence of antibody to human calicivirus in general population of Northern Japan, *J. Med. Virol.*, 7, 221, 1981.
42. **Cubitt, W. D. and McSwiggan, D. A.,** Calicivirus gastroenteritis in North West London, *Lancet*, 2, 975, 1981.
43. **Nakata, S., Chiba, S., Terashima, M., Sakuma, Y., Kogasaka, R., and Nakao, T.,** Microtitre solid-phase radioimmunoassay for detection of human calicivirus in stools, *J. Clin. Microbiol.*, 17, 198, 1983.
44. **Cubitt, W. D. and Barrett, A. D. T.,** Propagation of human candidate calicivirus in cell culture, *J. Gen. Virol.*, 65, 1123, 1984.
45. **Adler, I. and Zickl, R.,** Winter vomiting disease, *J. Infect. Dis.*, 119, 668, 1969.
46. **Dolin, R., Blacklow, N. R., Dupont, H., Formal, S., Buscho, R. F., Kasel, J. A., Chames, R. P., Hornick, R., and Chanock, R. M.,** Transmission of acute infectious nonbacterial gastroenteritis to volunteers by oral administration of stool filtrates, *J. Infect. Dis.*, 123, 307, 1971.
47. **Dolin, R., Blacklow, N. R., Dupont, H., Buscho, R. F., Wyatt, R. G., Kasel, J. A., Hornick, R., and Chanock, R. M.,** Biological properties of Norwalk agent of acute infectious nonbacterial gastroenteritis, *Proc. Soc. Exp. Biol. Med.*, 140, 578, 1972.
48. **Kapikian, A. Z., Grein, J. L., Wyatt, R. G., Thornhill, T. S., and Chanock, R. M.,** Density in cesium chloride of the 27 nm "SF 11a" particle associated with acute infectious nonbacterial gastroenteritis, determination by ultracentrifugation and immune electron microscopy, *Proc. Soc. Exp. Biol. Med.*, 142, 874, 1973.
49. **Thornhill, T. S., Kalica, A. R., Wyatt, R. G., Kapikian, A. Z., and Chanock, R. M.,** Pattern of shedding of the Norwalk particle in stools during experimentally induced gastroenteritis in volunteers as determined by immune electron microscopy, *J. Infect. Dis.*, 132, 28, 1975.
50. **Wyatt, R. G., Dolin, N. R., Blacklow, H., Dupont, L., Buscho, R. F., Thornhill, T. S., Kapikian, A. Z., and Chanock, R. M.,** Comparison of three agents of acute infectious nonbacterial gastroenteritis by cross-challenge in volunteers, *J. Infect. Dis.*, 129, 709, 1974.
51. **Schreiber, D. S., Blacklow, N. R., and Trier, J. S.,** The mucosal lesion of the proximal small intestine in acute infectious nonbacterial gastroenteritis, *N. Engl. J. Med.*, 288, 1318, 1973.
52. **Blacklow, N. R., Dolin, R., Fedson, D. S., DuPont, H., Northrup, R. S., Hornick, R. B., and Chanock, R. M.,** Acute infectious non bacterial gastroenteritis: etiology and pathogenesis. A combined clinical staff conference at the clinical center of the National Institutes of Health, *Ann. Intern. Med.*, 76, 993, 1972.
53. **Agus, S. G., Dolin, R., Wyatt, R. G., Tousimis, A. J., and Northrup, R. S.,** Acute infectious nonbacterial gastroenteritis: intestinal histopathology. Histologic and enzymatic alterations during illness produced by the Norwalk agent in man, *Ann. Intern. Med.*, 79, 18, 1973.
54. **Greenberg, H. B., Wyatt, R. G., Valdesuso, J., Kalica, A. R., London, W. T., Chanock, R. M., and Kapikian, A. Z.,** Solid-phase microtiter radioimmunoassay for detection of the Norwalk strain of acute nonbacterial epidemic gastroenteritis virus and its antibodies, *J. Med. Virol.*, 2, 97, 1978.
55. **Blacklow, N. R., Cukor, G., Bedigian, M. K., Escheverria, P., Greenberg, H. B., Schreiber, D. S., and Trier, J. S.,** Immunoresponse and prevalence of antibody to Norwalk enteritis virus as determined by radioimmunoassay, *J. Clin. Microbiol.*, 10, 903, 1979.
56. **Greenberg, H. B., Valdesuso, J., Kapikian, A. Z., Chanock, R. M., Wyatt, R. G., Szmuness, W., Larrick, J., Kaplan, J., Gilman, R. H., and Sack, D. A.,** Prevalence of antibody to the Norwalk virus in various countries, *Infect. Immun.*, 26, 270, 1979.
57. **Wyatt, R. G. and Kapikian, A. Z.,** Viral agents associated with acute gastroenteritis in humans, *Am. J. Clin. Nutr.*, 30, 1857, 1977.
58. **Wyatt, R. G., Greenberg, H. B., Dalgard, D. W., Allen, W. P., Sly, D. L., Thornhill, T. S., Chanock, R. M., and Kapikian, A. Z.,** Experimental infection of chimpanzees with the Norwalk agent of epidemic gastroenteritis, *J. Med. Virol.*, 2, 89, 1978.
59. **Appleton, H., Buckley, M., Thom, B. T., Cotton, J. L., and Henderson, S.,** Virus-like particles in winter vomiting disease, *Lancet*, 1, 409, 1977.
60. **Thornhill, T. S., Wyatt, R. G., Kalica, A. R., Dolin, R., Chanock, R. M., and Kapikian, A. Z.,** Detection by immune electron microscopy of 26-27 nm virus like particles associated with two family outbreaks of gastroenteritis, *J. Infect. Dis.*, 135, 20, 1977.
61. **Greenberg, H. B., Wyatt, R. G., and Kapikian, A. Z.,** Norwalk virus in vomitus, *Lancet*, 1, 55, 1979.
61a. **Curry, A.,** personal communication.

61b. **Caul, E. O.,** personal observations.

62. **Caul, E. O., Ashley, C. R., and Pether, J. V. S.,** Norwalk-like particles in epidemic gastroenteritis in the U.K., *Lancet*, 2, 1292, 1979.

63. **Curry, A. and Roberts, J. L.,** Realistic look at small, round faecal viruses found by electron microscopy, *Lancet*, 1, 99, 1980.

64. **Greenberg, H. B., Valdesuso, J., Kalica, A. R., Wyatt, R. G., McAuliffe, V. J., Kapikian, A. Z., and Chanock, R. M.,** Proteins of Norwalk virus, *J. Virol.*, 37, 994, 1981.

65. **Kapikian, A. Z., Greenberg, H. B., Wyatt, R. G., Kalica, A. R., and Chanock, R. M.,** The Norwalk group of viruses — agents associated with epidemic viral gastroenteritis, in *Virus Infections of the Gastrointestinal Tract*, Tyrrell, D. A. J. and Kapikian, A. Z., Eds., Marcel Decker, New York, 1982.

66. **Pether, J. V. S. and Caul, E. O.,** An outbreak of food-borne gastroenteritis in two hospitals associated with a Norwalk-like virus, *J. Hyg. Camb.*, 91, 343, 1983.

67. **Riordan, T., Craske, J., Roberts, J. L., and Curry, A.,** Food borne infection by Norwalk-like virus (small round structured virus), *J. Clin. Pathol.*, 37, 817, 1984.

68. **Ellis, M. E., Watson, B., Mandal, B. K., Dunbar, E. M., Craske, J., Curry, A., Roberts, J., and Lomax, J.,** Micro-organisms in gastroenteritis, *Arch. Dis. Child.*, 59, 848, 1984.

69. **Gill, O. N., Cubitt, W. D., McSwiggan, D. A., Watney, B. M., and Bartlett, C. L. R.,** Epidemic of gastroenteritis caused by oysters contaminated with small round structured viruses, *Br. Med. J.*, 287, 1532, 1983.

70. **Murphy, A. M., Grohmann, G. S., Christopher, P. J., Lopez, W. A., Davey, G. R., and Millson, R. H.,** An Australia-wide outbreak of gastroenteritis from oysters caused by Norwalk virus, *Med. J. Aust.*, 2, 329, 1979.

71. **Grohmann, G. S., Murphy, A. M., Christopher, P. J., Auty, E., and Greenberg, H. B.,** Norwalk virus gastroenteritis in volunteers consuming depurated oysters, *Aust. J. Exp. Biol. Med. Sci.*, 59, 219, 1981.

72. **Eyles, M. J., Davey, G. R., and Huntley, E. J.,** Demonstration of viral contamination of oysters responsible for an outbreak of viral gastroenteritis, *J. Food Prot.*, 44, 294, 1981.

73. **Appleton, H. and Pereira, M. S.,** A possible virus aetiology in outbreaks of food poisoning from cockles, *Lancet*, 1, 780, 1977.

74. **Linco, S. J. and Grohman, G. S.,** The Darwin outbreak of oyster-associated gastroenteritis, *Med. J. Aust.*, 1, 211, 1980.

74a. **Caul, E. O.,** unpublished observations.

75. **Appleton, H.,** Outbreaks of gastroenteritis associated with food, *De Ware (n) - Chemicus*, 13, 37, 1983.

76. **Caul, E. O., Ashley, C. R., and Egglestone, S. I.,** An improved method for the routine identification of faecal viruses using ammonium sulphate precipitation, *FEMS Microbiol. Lett.*, 4, 1, 1978.

77. **Taniguchi, K., Urasawa, S., and Urasawa, T.,** Virus-like particle, 35 to 40 nm, associated with an institutional outbreak of acute gastroenteritis in adults, *J. Clin. Microbiol.*, 10, 730, 1979.

78. **Taniguchi, K., Urasawa, S., and Urasawa, T.,** Further studies of 35-40 nm virus-like particles associated with outbreaks of acute gastroenteritis, *J. Med. Microbiol.*, 14, 107, 1981.

79. **Kogasaka, R., Nakamura, S., Chiba, S., Sakuma, Y., Terashima, H., Yokoyama, T., and Nakao, T.,** The 33-to-39 nm virus-like particles, tentatively designated as Sapporo agent, associated with an outbreak of acute gastroenteritis, *J. Med. Virol.*, 8, 187, 1981.

80. **Oshiro, L. S., Haley, C. E., Roberto, R. R., Riggs, J. L., Croughan, M., Greenberg, H., and Kapikian, A.,** A 27-nm virus isolated during an outbreak of acute infectious nonbacterial gastroenteritis in a convalescent hospital: a possible new serotype, *J. Infect. Dis.*, 143, 791, 1981.

81. **Dolin, R., Reichman, R. C., Roessner, K. D., Tralka, T. S., Schooley, R. T., Gary, W., and Morens, D.,** Detection by immune electron microscopy of the snow mountain agent of acute viral gastroenteritis, *J. Infect. Dis.*, 146, 184, 1982.

82. **Paver, W. K., Caul, E. O., and Clarke, S. K. R.,** Comparison of a 22 nm virus from human faeces with animal parvoviruses, *J. Gen. Virol.*, 22, 447, 1974.

83. **Paver, W. K. and Clarke, S. K. R.,** Comparison of human fecal and serum parvo-like viruses, *J. Clin. Microbiol.*, 4, 67, 1976.

84. **Cossart, Y. E., Field, A. M., Cant, B., and Widdows, D.,** Parvovirus-like particles in human sera, *Lancet*, 1, 72, 1975.

85. **Paver, W. K., Caul, E. O., and Clarke, S. K. R.,** Parvovirus-like particles in human sera, *Lancet*, 1, 232, 1975.

86. **Clarke, S. K. R., Cook, G. T., Egglestone, S. I., Hall, T. S., Miller, D. L., Reed, S. E., Rubenstein, D., Smith, A. J., and Tyrrell, D. A. J.,** A virus from epidemic vomiting disease, *Br. Med. J.*, 3, 86, 1972.

87. **Binn, L. N., Lazar, E. C., Eddy, J. A., and Kajima, M.,** Recovery and characterisation of a minute virus of canines, *Infect. Immun.*, 1, 503, 1970.
88. **Carmichael, L. E. and Binn, L. N.,** New enteric viruses in the dog, *Adv. Vet. Sci. Comp. Med.*, 25, 1, 1981.
89. **Christopher, P. J., Grohmann, G. S., Millsom, R. H., and Murphy, A. M.,** Parvovirus gastroenteritis. A new entity for Australia, *Med. J. Aust.*, 1, 121, 1978.

Chapter 11

ADENO-ASSOCIATED VIRUSES OF HUMANS

Neil R. Blacklow

TABLE OF CONTENTS

I. INTRODUCTION

The adeno-associated viruses (AAV) are small, DNA-containing agents that require coinfection with a helper virus for their own replication. AAV were originally discovered by electron microscopy as contaminants of laboratory stocks of tissue culture grown adenoviruses. Atchison and colleagues[1] first demonstrated their defective viral nature in their requirement for adenovirus for replication, even though AAV were unrelated immunologically and morphologically to adenovirus. AAV were subsequently recovered from human clinical specimens,[2] usually with adenovirus, and were the first example of human infection with a structurally independent yet defective virus. To date, AAV have not been associated conclusively with human disease. However, serological surveys indicate that human infections are common.[3] Because of their defectiveness and small genome, AAV have been widely studied by molecular biologists,[4] and it is fair to say that much more is known about the biochemical, biophysical, and immunological aspects of AAV at the molecular level than about their epidemiology and pathogenesis.

II. CLASSIFICATION AND GENERAL CHARACTERISTICS

AAV belong to the family Parvoviridae, and like other members possess single-stranded DNA. In 1981, the International Committee on Taxonomy of Viruses placed AAV within the newly designated genus *Dependovirus*[5] to indicate the helper virus dependency of AAV. This action was taken because recent studies indicated that, in contrast to earlier reports, the helper functions necessary for complete replication of AAV could be provided by coinfection with not only adenovirus but also herpes simplex virus.[6] However, individual members of the *Dependovirus* genus are still referred to as AAV-1 (i.e., AAV serotype 1) through AAV-5, and this terminology will be followed in this chapter.

AAV are 20 to 25 nm in diameter, are nonenveloped, and possess icosahedral symmetry. As is the case with other parvoviruses, AAV are highly heat resistant (60°C for at least 1 hr), stable at 0°C for months, and resistant to ether, chloroform, and anionic detergent.[7] AAV contain a small amount (1.55×10^6 daltons) of single-stranded linear DNA as their genome.[8] They also contain three polypeptides.[9] AAV virions have a high relative specific density in cesium chloride (1.38 to 1.44 g/cm^3).[10]

AAV are immunologically and morphologically distinct from their adenovirus and herpesvirus helpers. Five serotypes of AAV exist as demonstrated by serological studies.[3,10,11] AAV1, AAV2, AAV3, and AAV5 seem to infect humans as their major host, whereas AAV4 is a simian virus that rarely infects people. Nucleic acid hybridization studies show a 60 to 70% homology between AAV1, AAV2, AAV3, and AAV4.[5] Considerably less homology is seen between these four serotypes and the recently recognized AAV5.[11] Immunological studies such as complement fixation, immunodiffusion, neutralization, and immunofluorescence have been used to differentiate AAV types serologically. It is known that AAV types 1, 2, and 3 do not hemagglutinate erythrocytes, whereas AAV-4 has a hemagglutinin.[7]

AAV, in addition to the types 1 through 5 referred to above, also have been isolated from animal hosts other than humans. Bovine, canine, and avian AAV are officially recognized, and AAV from horses and sheep have been described.[5] It is interesting that AAV types 1 through 4 can often replicate in cells from an animal species other than human, provided that an adenovirus helper specific for the animal species is used as a coinfecting virus.[7,13] Thus, the human AAV serotypes are not tightly restricted in their host specificity unlike most human viruses. This observation reinforces currently unanswered questions about AAV: where do they come from and what is the clinical consequence of human infection with AAV?

III. BIOLOGY

The basic biology of AAV has been the subject of numerous studies (e.g., Reference 4). For the purposes of providing background information relevant for our subsequent discussion of AAV infection of humans, three features of AAV biology will be reviewed briefly, namely, replication, latency, and inhibition of their helper viruses.

A. Replication of AAV

AAV are completely defective viruses. If a cell is not coinfected with a helper virus, AAV will attach to the cell, penetrate the cellular membrane, enter the cytoplasm, and travel to the nucleus where the AAV genome is uncoated.[12] However, AAV-specific DNA, RNA, and protein synthesis will not occur as measured by traditional methods.[13] Addition of an adenovirus helper enables these events to take place with the release of mature infectious progeny AAV.

The small single-stranded DNA genome of AAV particles is comprised of complementary strands of DNA separately encapsidated so that approximately one half of viral particles contain a strand of positive polarity and the remainder of the virions contain a negative DNA strand. During AAV replication, plus and minus strands that infect the same cell combine to form a double-stranded intermediate.[14] Following this stage of viral replication, it is known that viral RNA transcription can only occur when cells are coinfected with a helper adenovirus. Resultant AAV mRNA is translated to form polypeptides of three size classes which comprise the AAV capsid.[9]

It was initially shown by several laboratories that herpes viruses only partially assisted in the replication of AAV in contrast to the complete helper functions provided by adenoviruses.[13,15-20] Coinfection of cells with AAV and a member of the herpes virus group (e.g., herpes simplex, cytomegalovirus, herpes zoster, Epstein Barr virus, and herpes virus saimiri) resulted in the production of AAV-specific proteins, RNA, and infectious nonencapsidated DNA but not mature infectious progeny. In 1981, Buller et al.[6] presented contradictory data indicating that herpes simplex virus can provide complete helper functions for AAV with between the 1981 report and previous studies exists at this time. In 1985, McPherson et al. reported that another herpes group virus, cytomegalovirus, completely helps AAV replication.[20a]

No other viruses seem capable of providing helper functions for AAV.

B. AAV Latency

It is now known that AAV is able to establish a latent state in infected cells in the absence of coinfection with a helper virus. This property of latency is not unique to AAV as it is shared by some other types of DNA viruses that infect cell nuclei. What is unique to AAV latency, however, is that the AAV genome is efficiently rescued and replicated when the cell is coinfected with a helper virus. Both in vivo and in vitro models for AAV latency exist which are now described briefly.

In vivo latent AAV infection has been discovered in primary human and monkey cell culture lots screened in the U.S. as part of a government vaccine production evaluation program to evaluate these cells for contamination with endogenous viruses.[12] AAV structural antigens were not detected in these cell cultures. However, when these cells were infected with adenoviruses, 1 to 2% of the cell lots of human origin and up to 20% of the simian cell lots released infectious AAV. Therefore, it is clear that primary cells used for vaccine production should be screened for the presence of contaminating AAV.

Latent AAV infection in vitro has been produced in a continuous line of human cells, Detroit 6, inoculated with AAV-2.[21] These cells seem to be biologically equivalent to the in vivo latent infection described above. Detroit 6 cells are negative for AAV antigens and infectivity until challenged by infection with helper adenovirus. The AAV DNA seems to be integrated into cellular DNA in these cells.[12] It is of interest that cells latently infected with AAV are not transformed in the usual sense and show no recognizable changes in their phenotypes.

C. Inhibition of Helper Viruses by AAV

It is paradoxical that, although AAV require helper viruses for their replication, they in turn exhibit inhibitory effects on their helpers. These inhibitory effects are manifested by a reduction in infectivity and antigen titers of helper viruses,[7,10] as well as an abrogation of their properties of cell transformation and oncogenicity.[22-25] Laboratory stocks of adenoviruses that lose titer on serial propagation should be suspected of being contaminated with AAV, which is a common laboratory phenomenon. AAV have also been shown in animal models to reduce the expression of tumorigenicity by adenovirus and herpes simplex virus. It is unclear whether this antitumor effect is due to a toxic action of AAV on the neoplastic cell or to interference by AAV with a function of the transforming helper virus.

IV. EPIDEMIOLOGY OF HUMAN INFECTION WITH AAV

The epidemiology of human infection with an unconditionally defective virus, AAV, is a very interesting topic that raises several questions. Does AAV infection require coinfection with a helper virus in humans as occurs in vitro? Which body sites are infected with AAV? How is this defective virus maintained in nature? Is human infection common or a rare event?

A. Distribution of Antibodies to AAV

Serum antibodies to AAV1, AAV2, and AAV3 have been detected frequently in humans by neutralization, complement fixation, and immunofluorescence tests. The age-related curve of antibody frequency shows a steep rise during childhood in the U.S.[3,27] which is also the peak age for adenovirus infections (Figure 1). About one half of children older than 10 years of age as well as adults possess antibodies to one or more of the first three AAV serotypes. Antibodies are most frequently found (in about one half of humans) against AAV2 and AAV3, which are also the two serotypes known to show the greatest degree of serological cross-relatedness. Antibodies are somewhat less often found against AAV1.

Seroconversions to AAV1, 2, and 3, that probably reflect broadly heterotypic antibody responses, have been shown in 12 of 40 children residing in a receiving home for young children during two different time periods.[3] These serological data provided the first evidence that man is naturally infected with AAV.

Antibody to AAV4 is rarely detected in human sera.[3] One exception is its reported presence in military recruits who received respiratory virus vaccines prepared in monkey kidney tissue cultures.[28] It is not known whether these recruits may have developed AAV4 antibody from infections due to activation of latent defective viruses in monkey kidney cells or to antigenic components in the vaccines themselves. Antibody to AAV4 is commonly found in sera from African green monkeys, and less often in rhesus monkeys. It is likely that AAV4 is a simian virus.

Serum antibody to AAV-5 is found by enzyme-immunoassay in approximately one half of adults in West Germany (Figure 2).[29] The sero-epidemiology of AAV-5 infection differs from that of AAV1, 2, and 3 in that the age-related curve of AAV-5 antibody frequency rises at later ages in a manner noted to be similar to that for antibody against herpes simplex antigens.[29] The highest antibody titers against AAV-5 are found in the age group between 15 and 20 years.

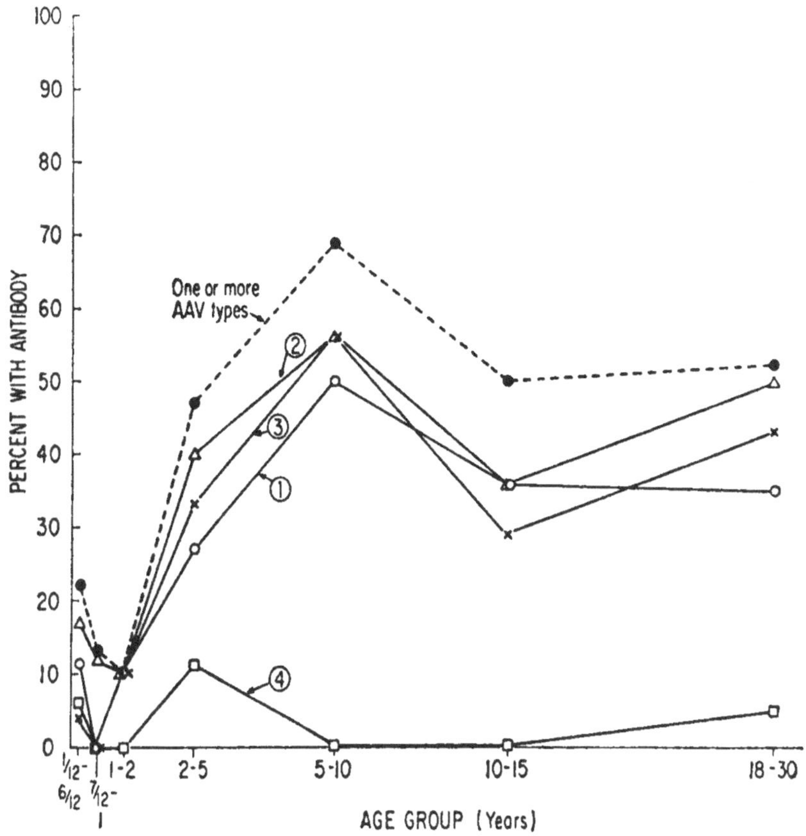

FIGURE 1. Age distribution of AAV neutralizing antibody. AAV serotype is indicated by the *circled* number. (From Blacklow, N. R., Hoggan, M. D., and Rowe, W. P., *J. Natl. Cancer Inst.*, 40, 319, 1968. Reproduced by courtesy of the editor of the *Journal of the National Cancer Institute*. With permission.)

B. Isolation of AAV From Humans

Conclusive proof that AAV infect humans is provided by the isolation of AAV types 2, 3, and 5 from human clinical specimens.[2,29,30] AAV have been cultivated in vitro from clinical materials only when cell cultures are coinfected with a helper adenovirus. Stringent controls have been used in these studies to rule out possible AAV contaminants of cell cultures and of helper adenoviruses.

AAV2 and AAV3 were first isolated from anal swab specimens from 23 children in a Washington, D.C. receiving home. A total of 11 of 12 children tested subsequently developed antibody to AAV.[2,30] AAV2 was also isolated from six throat swab specimens. The isolation of both AAV2 and AAV3 from these children indicates that both viruses circulated in this population simultaneously. All 36 AAV isolations were made from 23 children experiencing acute adenovirus infection, which occurred with ten different adenovirus serotypes. AAV isolates possessed properties identical to the prototype laboratory strains.

Later, 35 strains of AAV3 were isolated from pediatric and adult patients with swimming pool-associated conjunctivitis due to adenovirus type 3 in Seattle.[31] Twenty-one of the AAV3 isolates were collected from conjunctival swabs, eight from throat swabs, and six from fecal samples.

AAV-5 was originally discovered by its isolation from a male with a penile flat condylomatous lesion.[11] Material from this lesion was inoculated into human fibroblast cultures which were coinfected with a helper adenovirus. No other isolations of AAV-5 are reported

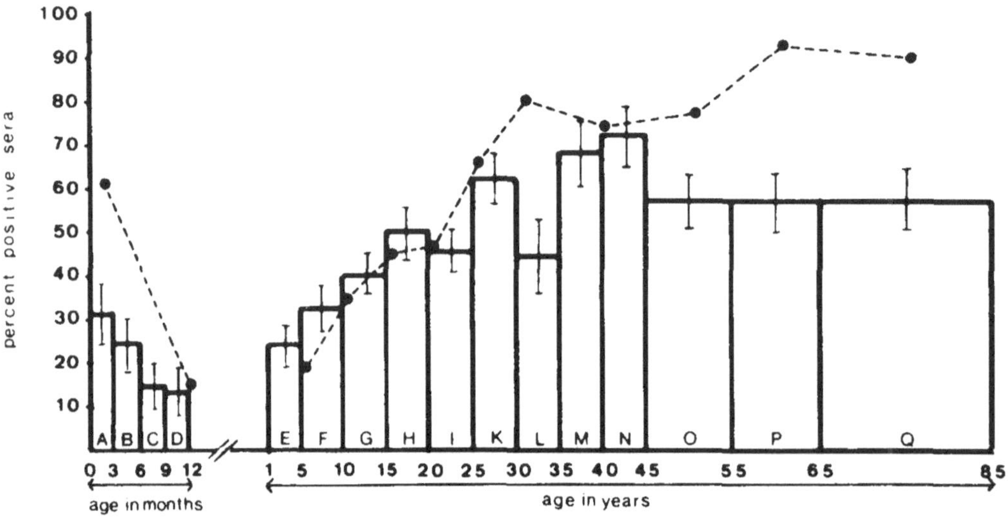

age group	A	B	C	D	E	F	G	H	I	K	L	M	N	O	P	Q	total
n	51	54	42	38	59	61	67	60	92	69	50	41	54	68	52	54	912
% ♂	47.1	61.1	54.8	71.1	72.9	60.7	58.2	48.3	40.2	37.7	44.0	41.5	46.3	55.9	55.8	48.1	437
% ♀	52.9	38.9	42.9	28.9	27.1	39.3	41.8	51.7	59.8	60.9	56.0	58.5	51.9	44.1	44.2	51.9	475

FIGURE 2. Age-specific occurrence of antibodies to AAV5. The bars indicate confidence limits. The dotted lines reveal the percentage of herpes simplex virus-positive sera as determined within the same population. The insert indicates the number of sera per age group as well as the percentages of sera from male and female patients. (From Georg-Fries, B., Biederlack, S., Wolf, J., and ZurHausen, H., *Virology*, 134, 64, 1984. Reproduced by courtesy of Academic Press. With permission.)

to date. Unlike the other AAV serotypes which were originally discovered as laboratory contaminants of adenovirus preparations, AAV-5 has yet to be reported as an endogenous agent in cell cultures.

C. Mechanism of Spread of AAV

The person to person transmission of AAV infection is novel in that the virus appears to travel with its helper adenovirus. This observation is based mostly on an epidemiological study of AAV2 and AAV3 infection during an 8-week period in a nursery population.[30] In this population, adenovirus was isolated from an individual on the same date as 34 of the 36 AAV isolates. In the two remaining instances, adenovirus was isolated from the children 2 weeks prior to AAV. These data show an intimate temporal association of adenovirus and AAV infections in humans. In addition, the distribution of coinfecting adenovirus serotypes from AAV-negative children was approximately the same as that among AAV-positive children, reinforcing the fact that multiple adenovirus serotypes can serve as helpers for AAV in humans without a specific preference for any one adenovirus serotype. Finally, AAV tended to be isolated from children who had more prolonged adenovirus infections with extended viral shedding; proionged adenovirus infection probably provides an ideal lawn for AAV replication either as an initial infection or as an efficient reactivation of AAV genome from latency.

V. POSSIBLE DISEASE ASSOCIATIONS

AAV have yet to be associated conclusively with human disease. Speculation has centered on possible disease manifestations related to the important features of the biology of AAV which were discussed above in Section III. These include their defectiveness with reliance on adenovirus or herpesvirus helpers, their properties of latency, and their ability to exert inhibitory effects on their helpers.

A. Effects on Acute Adenovirus Infections

Outbreaks of AAV infection, coincident with adenovirus, have not been associated with any unique disease manifestations. The epidemiological study cited above of AAV infections during an 8-week period in a nursery population failed to show a change in numbers of febrile illnesses in AAV infected children when they were compared with AAV-negative controls.[30] Also, there appeared to be no increase in diarrhea, rash, or unusual illness in AAV-infected children.

Subsequently, a large sero-epidemiological study of AAV infection was performed to assess the possible participation of AAV in the disease response to acute adenovirus infection requiring hospitalization.[32] A search for complement-fixing antibodies to AAV2 and AAV3 was conducted on paired sera of 551 hospitalized children as part of an 8-year controlled study of adenovirus respiratory illness. Patients were studied in three categories: (1) those with respiratory tract disease associated with adenovirus; (2) adenovirus-negative asymptomatic controls; and (3) adenovirus-infected asymptomatic controls. Approximately 30% of individuals in each category possessed antibodies to AAV. A total of 31 of 32 children with seroconversions to AAV had evidence of a recent adenovirus infection. Of the 32 children who seroconverted to AAV, 30 were in the category of adenovirus-infected patients ill with respiratory tract disease. Pneumonia, bronchiolitis, croup, and pharyngitis-bronchitis occurred at approximately the same frequency among children with an AAV seroconversion as among all persons with respiratory illness. Finally, patients with an AAV seroconversion tended to show the virologic and serologic features of adenovirus infections associated with respiratory tract disease. Thus, there was no evidence that AAV inhibited the disease manifestations of adenovirus infection and no indication that AAV enhanced adenovirus pathogenicity.

One study in Seattle has reported a high isolation rate of AAV3 from patients involved in a swimming pool-associated conjunctivitis outbreak due to adenovirus type 3.[31] AAV3 was isolated more frequently from conjunctival specimens than from throat or fecal samples. In contrast, AAV3 was infrequently found with adenoviruses isolated from other population groups in Seattle. The swimming pool conjunctivitis report with its high isolation rate of AAV3 may merely be an incidental association reflecting a common source infection. Alternatively, AAV might play a role in conjunctivitis.

B. Malignancy

As mentioned earlier, AAV are known in animal models to reduce the expression of tumorigenicity by their two helper viruses, adenovirus and herpes simplex virus.[22,24,25] There are also clues from a few sero-epidemiological studies in humans that AAV infection could conceivably have an inhibitory effect on oncogenesis. However, the implications of these studies are entirely theoretical because they contain the assumption that adenovirus or herpes simplex virus cause malignancies in humans, which is, of course, not established.

One sero-epidemiological study in Belgium revealed that AAV3 antibodies were found in 15% of 337 patients with a variety of tumors compared with 34% of 268 control individuals.[33] No such differences between groups were noted in antibodies to AAV2, adenovirus, herpes simplex, or cytomegalovirus. When types of cancer were evaluated, 46 patients with cervical cancer significantly differed from their matched controls by a high prevalence of adenovirus antibodies without AAV3 antibodies. In contrast, a high prevalence of all cancer patients had combinations of antibodies to AAV2 and adenovirus, and also to AAV3 and cytomegalovirus. Thus, in cancer patients, the epidemiology of AAV2 and AAV3 differed.

A report from Texas evaluated antibodies to AAV2, AAV3, and herpes simplex virus in sera from cancer patients and normal adults.[34] Interestingly, a markedly lower incidence of AAV antibody, 14%, was found among 70 cervical cancer patients (a tumor possibly associated with herpes simplex virus infection) than among a matched control group. The latter demonstrated the expected 80% of AAV antibody prevalence in that study. Meanwhile, antibody prevalences to herpes simplex virus were similar in both the cervical cancer and control groups. This report is compatible with the concept that women with evidence of prior herpes simplex virus infection but without evidence of prior AAV infection have an increased risk of developing cervical cancer.

In view of the two preceding reports[33,34] showing a low prevalence of AAV antibodies in cervical cancer patients, a similar pilot serological study was carried out with AAV5, the most recently recognized AAV serotype which was discovered, interestingly, in a genital lesion.[29] Of 83 cervical cancer patients, 33 (39.7%) lacked AAV5 antibodies by ELISA, and of 50 matched controls, 16 (32%) were devoid of AAV5 antibodies. However, geometric mean titers of positive sera were 216 in the cancer group and 636 in the control group. Thus, control patients had about threefold higher antibody titers against AAV5 than did cervical cancer patients. These results are of uncertain significance because the authors did not provide a detailed epidemiological analysis of their study groups. For example, socioeconomic factors might explain their findings. At present the studies of Sprecher-Goldberger,[33] Mayor,[34] and Georg-Fries[29] must be viewed cautiously. However, they have raised a specific question: do infections by AAV inhibit subsequent development of cervical cancer?

C. Congenital Infections

Many autonomous parvoviruses cause fetal infections in animals, with resultant stillbirths and congenital malformations. Much less information is available about AAV in this regard. In a murine model, it is known that AAV1 will produce transplacental infection in the presence of coinfection with helper murine adenovirus.[35] AAV can be detected in the kidneys, livers, and lungs of fetuses and newborns after their pregnant mothers have been inoculated with AAV and adenovirus. Abnormalities attributed to AAV in this murine model include retardation of fetal development, smaller newborn litter size, and reduction in average weight of newborns.

These studies with the murine model have raised the question of AAV persistence or latency in humans. Certainly, models for in vitro latency in Detroit 6 and KB cells exist[21,36] in which infectious AAV can be rescued from latency by infection with helper virus. Furthermore, 1 to 2% of human embryos are known to be covertly infected with AAV2.[12] In an attempt to follow a human model as closely as possible, fetuses of eight pregnant rhesus monkeys were inoculated in utero with AAV1 and/or simian adenovirus SV-15 helper.[12] It was not possible to demonstrate any AAV latency as fetuses inoculated with AAV alone did not show AAV antibody even after SV15 challenge post-delivery. Adenovirus was necessary for AAV replication in this system as AAV was isolated from spontaneously aborted fetal tissues only when fetuses had been coinfected with pathogenic SV-15 virus.

One serological study has been performed to determine if AAV infection in humans correlates with reproductive failure or teratogenic effects.[12] Eighty paired acute and convalescent serum samples were obtained from mothers with histories of reproductive failure, stillbirth, or spontaneous abortion. No significant trends were found in the frequencies or antibody titers to AAV1, AAV2, and AAV3. Thus, to date, no evidence exists for a role of AAV in human congenital infections.

REFERENCES

1. **Atchison, R. W., Casto, B. C., and Hammon, W. McD.,** Adenovirus-associated defective virus particles, *Science*, 194, 754, 1965.
2. **Blacklow, N. R., Hoggan, M. D., and Rowe, W. P.,** Isolation of adenovirus-associated viruses from man, *Proc. Natl. Acad. Sci. U.S.A.*, 58, 1410, 1967.
3. **Blacklow, N. R., Hoggan, M. D., and Rowe, W. P.,** Serologic evidence for human infection with adenovirus-associated viruses, *J. Natl. Cancer Inst.*, 40, 319, 1968.
4. **Ward, D. C. and Tattersall, P., Eds.,** *Replication of Mammalian Parvoviruses*, Cold Spring Harbor Laboratory, Cold Spring Harbor, N.Y., 1978.
5. **Siegl, G., Bates, R. C., Berns, K. I., Carter, B. J., Kelly, D. C., Kurstak, E., and Tattersall, P.,** Characteristics and taxonomy of Parvoviridae, *Intervirology*, 23, 61, 1985.
6. **Buller, R. M. L., Janik, J. E., Sebring, E. D., and Rose, J. A.,** Herpes simplex virus type 1 and type 2 completely help adenovirus-associated virus replication, *J. Virol.*, 40, 241, 1981.
7. **Hoggan, M. D.,** Adenovirus-associated viruses, in, *Progress in Medical Virology*, Vol. 12, Melnick, J. L., Ed., S. Karger, Basel, 1970, 211.
8. **Gerry, H. W., Kelly, T. J., Jr., and Berns, K. I.,** Arrangement of nucleotide sequences in adeno-associated virus DNA, *J. Molec. Biol.*, 79, 207, 1973.
9. **Johnson, F. B., Blacklow, N. R., and Hoggan, M. D.,** Immunological reactivity of antisera prepared against the sodium dodecyl sulfate-treated structural polypeptides of adenovirus-associated virus, *J. Virol.*, 9, 1017, 1972.
10. **Hoggan, M. D., Blacklow, N. R., and Rowe, W. P.,** Studies of small DNA viruses found in various adenovirus preparations: physical, biological and immunological characteristics, *Proc. Natl. Acad. Sci. U.S.A.*, 55, 1467, 1966.
11. **Bantel-Schaal, U. and ZurHausen, H.,** Characterization of the DNA of a defective human parvovirus isolated from a genital site, *Virology*, 134, 52, 1984.
12. **Cukor, G., Blacklow, N. R., Hoggan, M. D., and Berns, K. I.,** Biology of adeno-associated virus in *Replication of Mammalian Parvoviruses*, Ward, D. C. and Tattersall, P., Eds., Cold Spring Harbor Laboratory, Cold Spring Harbor, N.Y., 1978, 33.
13. **Rose, J. A. and Koczot, F.,** Adenovirus-associated virus multiplication. VII. Helper requirement for viral deoxyribonucleic acid and ribonucleic acid synthesis, *J. Virol.*, 10, 1, 1972.
14. **Straus, S. E., Sebring, E. D., and Rose, J. A.,** Concatemers of alternating plus and minus strands are intermediates in adenovirus-associated virus DNA synthesis, *Proc. Natl. Acad. Sci. U.S.A.*, 73, 742, 1976.
15. **Atchison, R. W.,** The role of herpesvirus in adenovirus-associated virus replication in vitro, *Virology*, 42, 155, 1970.
16. **Blacklow, N. R., Hoggan, M. D., and McClanahan, M. S.,** Adenovirus-associated viruses: enhancement by human herpesviruses, *Proc. Soc. Exp. Biol. Med.*, 134, 952, 1970.
17. **Boucher, D. W., Melnick, J. L., and Mayor, H. D.,** Nonencapsidated infectious DNA of adeno-satellite virus in cells coinfected with herpesvirus, *Science*, 173, 1243, 1971.
18. **Henry, C. J., Merkow, L. P., Pardo, M., and McCabe, C.,** Electron microscope study on the replication of AAV-1 in herpes-infected cells, *Virology*, 49, 618, 1972.
19. **Dolin, R. and Rabson, A. S.,** Herpesvirus saimiri: enhancement of adenovirus-associated virus, *J. Natl. Cancer Inst.*, 50, 205, 1973.
20. **Young, J. F. and Mayor, H. D.,** Studies on the defectiveness of adeno-associated virus (AAV). II. Effect of growth in CV-1 cells on the replication of adeno-associated virus, *Virology*, 94, 342, 1979.

20a. **McPherson, R. A., Rosenthal, L. J., and Rose, J. A.,** Human cytomegalovirus completely helps adeno-associated virus replication, *Virology*, 147, 217, 1985.

21. **Berns, K. I., Pinkerton, T. C., Thomas, G. F., and Hoggan, M. D.,** Detection of adeno-associated virus (AAV)-specific nucleotide sequences in DNA isolated from latently infected Detroit 6 cells, *Virology*, 68, 556, 1975.
22. **Kirschstein, R. L., Smith, K. O., and Peters, E. A.,** Inhibition of adenovirus 12 oncogenicity by adeno-associated virus, *Proc. Soc. Exp. Biol. Med.*, 128, 670, 1968.
23. **Casto, B. C. and Goodheart, C. R.,** Inhibition of adenovirus transformation in vitro by AAV-1, *Proc. Soc. Exp. Biol. Med.*, 140, 72, 1972.
24. **Mayor, H. D., Houlditch, G. S., and Mumford, D. M.,** Influence of adeno-associated satellite virus on adenovirus-induced tumors in hamsters, *Nature New Biol.*, 241, 44, 1973.
25. **Cukor, G., Blacklow, N. R., Kibrick, S., and Swan, I. C.,** Effect of adeno-associated virus on cancer expression by herpesvirus-transformed hamster cells, *J. Natl. Cancer Inst.*, 55, 957, 1975.
26. **Blacklow, N. R., Cukor, G., Kibrick, S., and Quinnan, G.,** Interactions of adeno-associated viruses with cells transformed by herpes simplex virus in, *Replication of Mammalian Parvoviruses*, Ward, D. C. and Tattersall, P., Eds., Cold Spring Harbor Laboratory, Cold Spring Harbor, N.Y., 1978, 87.

27. **Parks, W. P., Boucher, D. W., Melnick, J. L., Taber, H., and Yow, M. D.,** Seroepidemiological and ecological studies of the adenovirus-associated satellite viruses, *Infect. Immun.*, 2, 716, 1970.
28. **Rosenbaum, M. J., Edwards, E. A., Pierce, W. E., Peckinpaugh, R. O., Parks, W. P., and Melnick, J. L.,** Serologic surveillance for adeno-associated satellite virus antibody in military recruits, *J. Immunol.*, 106, 711, 1971.
29. **Georg-Fries, B., Biederlack, S., Wolf, J., and ZurHausen, H.,** Analysis of proteins, helper dependence, and seroepidemiology of a new human parvovirus, *Virology*, 134, 64, 1984.
30. **Blacklow, N. R., Hoggan, M. D., Kapikian, A. Z., Austin, J. B., and Rowe, W. P.,** Epidemiology of adenovirus-associated virus infection in a nursery population, *Am. J. Epidemiol.*, 8, 368, 1968.
31. **Schmidt, O. W., Cooney, M. K., and Foy, H. M.,** Adeno-associated virus in adenovirus type 3 conjunctivitis, *Infect. Immun.*, 11, 1362, 1975.
32. **Blacklow, N. R., Hoggan, M. D., Sereno, M. S., Brandt, C. P., Kim, H. W., Parrott, R. H., and Chanock, R. M.,** Seroepidemiologic study of adeno-associated virus infection in infants and children, *Am. J. Epidemiol.*, 94, 359, 1971.
33. **Sprecher-Goldberger, S., Thiry, L., Lefebvere, N., Dekegel, D., and DeHalleux, F.,** Complement-fixation antibodies to adenovirus-associated viruses, adenoviruses, cytomegaloviruses and herpes simplex viruses in patients with tumors and in control individuals, *Am. J. Epidemiol.*, 94, 351, 1971.
34. **Mayor, H. D., Drake, S., Stahmann, J., and Mumford, D. M.,** Antibodies to adeno-associated satellite virus and herpes simplex in sera from cancer patients and normal adults, *Am. J. Obstet. Gynecol.*, 126, 100, 1976.
35. **Lipps, B. V. and Mayor, H. D.,** Transplacental infection with adeno-associated virus type 1 in mice, *Intervirology*, 14, 118, 1980.
36. **Handa, H., Shiroki, K., and Shimojo, H.,** Establishment and characterization of KB cell lines latently infected with adeno-associated virus type 1, *Virology*, 82, 84, 1977.

INDEX

A

B

C

D

E

F

G

M

N

O

P

R

S

T

U

V

W